Routledge Revivals

Rural Development and Urban-Bound Migration in Mexico

Rapid growth of urban populations is a major characteristic of economic development and demographic change in developing countries leading to industrialisation and modernisation of major cities. Originally published in 1980, this study focusses on these issues using Mexico as a case study as well as analysing the risk of over-urbanisation and what the effects will be on cities such as Mexico City. This title will be of interest to students of Environmental studies and Economics.

Rural Development and Urban-Bound Migration in Mexico

Arthur Silvers and Pierre Crosson

First published in 1980
by Resources for the Future, Inc.

This edition first published in 2016 by Routledge
2 Park Square, Milton Park, Abingdon, Oxon, OX14 4RN
and by Routledge
605 Third Avenue, New York, NY 10017

Routledge is an imprint of the Taylor & Francis Group, an informa business

Publisher's Note
The publisher has gone to great lengths to ensure the quality of this reprint but points out that some imperfections in the original copies may be apparent.

Disclaimer
The publisher has made every effort to trace copyright holders and welcomes correspondence from those they have been unable to contact.

A Library of Congress record exists under LC control number: 80008024

ISBN 13: 978-1-138-19556-1 (hbk)
ISBN 13: 978-1-315-63832-4 (ebk)
ISBN 13: 978-1-138-19558-5 (pbk)

DOI: 10.4324/9781315638324

Rural development and urban-bound migration in Mexico

ARTHUR SILVERS and PIERRE CROSSON

RESEARCH PAPER R-17

RESOURCES FOR THE FUTURE / WASHINGTON, D.C.

RESOURCES FOR THE FUTURE, INC.
1755 Massachusetts Avenue, N.W., Washington, D.C. 20036

Resources for the Future is a nonprofit organization for research and education in the development, conservation, and use of natural resources and the improvement of the quality of the environment. It was established in 1952 with the cooperation of the Ford Foundation. Grants for research are accepted from government and private sources only if they meet the conditions of a policy established by the Board of Directors of Resources for the Future. The policy states that RFF shall be solely responsible for the conduct of the research and free to make the research results available to the public. Part of the work of Resources for the Future is carried out by its resident staff; part is supported by grants to universities and other nonprofit organizations. Unless otherwise stated, interpretations and conclusions in RFF publications are those of the authors; the organization takes responsibility for the selection of significant subjects for study, the competence of the researchers, and their freedom of inquiry.

Research Papers are studies and conference reports published by Resources for the Future from the authors' typescripts. The accuracy of the material is the responsibility of the authors and the material is not given the usual editorial review by RFF. The Research Paper series is intended to provide inexpensive and prompt distribution of research that is likely to have a shorter shelf life or to reach a smaller audience than RFF books.

Library of Congress Catalog Card Number 80-8024

ISBN 0-8018-2493-1

Manufactured in the United States of America

Published July 1980. $6.95.

Contents

		Page
Acknowledgments		viii
Chapter 1	INTRODUCTION	1
	Urbanization in Developing Countries	1
	Overview of the Research Design	6
	The econometric analysis	7
	The case study for the state of Sonora	12
	Layout of the Study	13
Chapter 2	ASPECTS OF URBAN POPULATION GROWTH IN MEXICO	15
	Dimensions of Urban Population Growth	15
	Sources of Urban Population Growth	20
	Natural increase and net migration	21
	Some Characteristics of Urban Migrants	24
Chapter 3	DETERMINANTS OF AGRICULTURAL INCOME	38
	Specification	38
	Definition and Measurement of Variables	39
	Means, Coefficients of Variation and Correlation Matrix	42
	Hypothesis Tests Via Multiple Regression	44
	Effects on State Agricultural Income of Out-Migration and Urban Wages	47
	Hypothesis Testing of Irrigation Impacts With Simple Regression	53
	Appendix to Chapter 3	56

Contents (cont.)

		Page
Chapter 4	DETERMINANTS OF RURAL TO URBAN MIGRATION	58
	Specification of the Migration Equation	58
	Definition and Measurement of Variables	60
	Estimation Problems	68
	Presentation and Discussion of the Data	69
	The Basic Migration Model	75
	First Variation on the Basic Migration Model	78
	Second Variation on the Basic Migration Model	82
	Third Variation on the Basic Migration Model	84
	Fourth Variation on the Basic Migration Model	87
	Next Steps	89
	Appendix to Chapter 4	91
Chapter 5	DETERMINANTS OF THE URBAN WAGE	97
	Introduction	97
	Specification	98
	Estimation Problems	103
	Definition and Measurement of Variables	105
	Hypothesis Tests Via Multiple Regression	106
Chapter 6	IMPLICATIONS FOR AN URBANIZATION POLICY	113
	Introduction	113
	An Appraisal of Policy Instruments	116
	Feasibility of Expanding Irrigation	120
	Systematic Urbanization Effects of Irrigation Development	122

Contents (cont.)

		Page
Chapter 7	IMPACT OF IRRIGATION INVESTMENTS ON AGRICULTURAL AND URBAN DEVELOPMENT IN SONORA	129
	Introduction	129
	Irrigation and Development in Sonora	131
	Conclusion	143
Chapter 8	CONCLUSION	145
Appendix	SOURCES OF DATA USED IN CHAPTERS 3-5	149

Diagrams

		Page
Diagram 1.	Block Recursive Structure	10
Diagram 2.	Simultaneous Structure	12
Diagram 3.	The Urbanization Effects of Rural Public Investment	125

Tables

		Page
Table 1.	Population Growth in Mexico, by Decade, 1900-1970	16
Table 2.	Primacy Indexes for Mexico	18
Table 3.	Shares of Selected Cities or Urban Areas in Mexico's Urban Population	19
Table 4.	Contribution of Net Migration to Population Growth in Urban Areas, 1940-1970	22
Table 5.	Age and Sex Distribution of Net Interstate Migration and of Total Population in Mexico, and Male-Female Ratio of Migrants	26
Table 6.	Mexico City: Distribution of Economically Active Males According to Occupation, by Migratory Status and Age, 1960	29
Table 7.	Monterrey: Socioeconomic Differences Between Native and Migrant Males, 21 to 60, 1965	30
Table 8.	Unemployment Rates of Natives and Migrants in Greater Santiago, 1963	34
Table 9.	Means and Coefficients of Variation of Variables Associated with State Agricultural Income per Worker, 1960	42
Table 10.	Correlation Matrix of Variables Assumed Linearly Associated with State Agricultural Income per Worker, 1960	43
Table 11.	Least Squares Regressions of 1960 Agricultural Income per Worker Against 1960 Values of Independent Variables, Log and Linear, All States	44
Table 12.	Simple Linear Correlations of OMRLF and AUWAG with Variables Specified in the AGINC Regression	49
Table 13.	Two-Stage Least Squares Log-Linear Regressions of Agricultural Income per Worker on Independent Variables	50

Table 14. Simple Linear Regression of Agricultural Income per Man on Irrigated Land per Man in 1960 55

Table 15. OLS Regression of 1950-1960 Growth Rate of AGINC Against 1950-1960 Growth Rates of Independent Variables 57

Table 16. Means and Coefficients of Variation of Variables Associated with State-to-City Migration 70

Table 17. Correlation Matrix of Variable Assumed Linearly Associated with Migration from 1960 to 1970 74

Table 18. State to City Migration, 1960 to 1970: the Basic Model 76

Table 19. State to City Migration, 1960 to 1970: First Variation on the Basic Model 80

Table 20. State to City Migration, 1960 to 1970: Second Variation on the Basic Model 82

Table 21. State to City Migration, 1960 to 1970: Third Variation on the Basic Model 84

Table 22. State to City Migration, 1960 to 1970: Fourth Variation on the Basic Model 88

Table 23. Simple Correlation Between 1950 and 1960 Values of Nine Correlates of Migration 94

Table 24. Distributed Lag Specification of the Log Form of the Migration Regression, $\lambda=.5$ 96

Table 25. Correlation Matrix of Variables Assumed Linearly Correlated with Urban Wage, 1965 107

Table 26. Average Urban Wage 1965, Estimated via Linear and Log Regression 108

Table 27. Attributes of the "Typical" Rural State and of Mexico City Used for the Impact Analysis 127

Table 28. Agricultural Production and Population: Region of Hermosillo and Ciudad Obregón 137

Table 29. Share of Cotton and Wheat in Total Value of Crop Production, Rio Yaqui and Costa de Hermosillo Irrigation Districts 138

Table 30. Fertilizer Use and Mechanization in Irrigation Districts of Mexico, by Region 140

Acknowledgments

We, like all students of migration and urbanization in Mexico, are deeply indebted to Luis Unikel and his colleagues at El Colegio de Mexico. Without their pioneering work in collecting and analyzing data relating to a broad spectrum of migration and urbanization issues in Mexico this study literally could not have been done. We hope we have achieved the high standard of scholarship they have established in this area of work.

We are grateful to Joel Bergsman, Emery Castle, Edgar Dunn, Michael Greenwood, Kenneth Frederick, Herbert Morton, and Ronald Ridker for careful readings of the manuscript and many useful comments, most of which we have tried to take into account. If defects remain, the fault is ours.

Arthur Silvers

Pierre Crosson

May, 1980

Chapter 1

INTRODUCTION

Urbanization in Developing Countries

Rapid growth of urban populations is one of the most prominent characteristics of economic development and demographic change in the developing countries. Since about 1940 urban populations in these countries have been growing by four to six percent annually, rates which double population about every 12 to 18 years. Much of the growth has been concentrated in one or two principal cities, some of which--for example, Mexico City, São Paulo, Rio de Janeiro, Buenos Aires, Calcutta--now are among the largest in the world, and still growing at rapid, if somewhat diminished, rates.

Attitudes of students of the process of urbanization and of policy makers concerned with it have varied over the last several decades.[1] For roughly ten years after the end of World War II, the dominant view was

---

[1]This discussion of attitudes toward urbanization is based on Alan Gilbert, Latin American Development: A Geographical Perspective (Penguin Books, 1974) pp. 83-127. Gilbert's account includes numerous references to the literature and the reader is referred to him for documentation of the argument made here. Debate on the advantages and disadvantages of urbanization did not begin after World War II, the period considered here. In modern times the debate goes back at least to the 18th century when the enclosure movement in England and the related rise of the industrial system forced people off the land and into the city where for many of them life was, in Hobbesian terms, nasty, brutish, and short. (See Karl Polanyi, The Great Transformation (Beacon Press, 1944) especially pp. 98-99, for an account of this period and references to contemporary writers on the social ills of early industrialism, both in the towns and in the countryside.) For a review of the debate about urbanization over a longer time span than that treated here, see B. J. L. Berry, The Human Consequences of Urbanization (New York, St. Martins, 1973).

that urbanization was a necessary and desirable concomitant of economic development. It was associated with industrialization, then the goal of virtually every developing country, and with the modernization of attitudes, economic practices, and social institutions of societies still locked in traditional modes. In the 1950s, however, this favorable view of urbanization in the developing countries was increasingly challenged by authors who pointed to the rapid emergence of urban slums, high unemployment, and associated problems of public health, crime, pollution of air and water, congestion, emotional stress, and erosion of morale among a significant underclass of the citizenry. These ills were attributed to "over-urbanization," an index of which was the ratio of service employment to total employment in urban areas. It was observed that this ratio was higher in most developing countries than it had been in developed countries at comparable levels of per capita income, and it was argued that this high ratio reflected more rapid migration to the cities than could be accommodated by industrial job creation. The result was high unemployment and the other stigmata of "over-urbanization."

By the late 1960s arguments for a more favorable view of urbanization once again became prominent. The "over-urbanization" hypothesis was criticized as giving a frequently distorted and sometimes false account of the consequences of urbanization. It was pointed out that much of the growth in service employment was in trade, finance, real estate, professional services, and government, activities generating relatively high levels of income. Thus the suggestion in the "over-urbanization" literature that most of the growth in service employment was among shoeshine boys, lottery salesmen, and other itinerant street vendors gave a quite incomplete account of the dynamics of the urbanization process. A study of Mexico, of particular interest here, showed that during the period of most rapid urban growth in Mexico, the expansion of industrial employment was more rapid than in service employment, the reverse of the relationship asserted by the "over-urbanization" hypothesis. The hypothesis also was criticized for suggesting that most urban growth was in one or two primate cities in each country, even though the evidence demonstrated that rapid growth occurred also in smaller cities. This clearly was the case in Mexico (see table 3 in the next chapter).

In our judgment the evidence is overwhelming that the net result of urbanization has been strongly favorable to the people of the developing countries. Economic history and theory both demonstrate that economic development requires an increasing specialization of labor. This implies a relative shift of labor and other resources from primary activities, principally agriculture, to industry, service, and other nonprimary lines of work. Since these other activities are much more sparing in the use of land than agriculture, their expansion relative to agriculture inevitably leads to increasing spatial concentrations of people, that is, to urban growth.

This favorable assessment of urbanization does not imply that agriculture should or can be neglected in the development process. On the contrary, we shall argue later in this report that the modernization of agriculture can contribute not only to higher incomes of farm people but also the growth of income and population in urban areas serving them. Nor does this assessment overlook the private and social costs of rapid urbanization. These costs are real and frequently high. The oft-cited air pollution of Mexico City, for example, is a fact, and anyone who has experienced it cannot deny that it exacts real social costs. The same can be said of the congestion and many of the other consequences of fast urbanization.

The favorable judgment of the consequences of urbanization so far experienced is not inconsistent with the view that at some future date the social costs of the process may outweigh the gains. For many urban areas in the developing countries, that date may be near, and some may already have passed beyond it. This possibility, if not likelihood, in fact is the rationale for this study. While we believe that on balance the people of Mexico have benefited strongly from urbanization, the costs of additional growth in some areas, Mexico City, for example, may soon exceed the additional benefits. The Mexican government in any case apparently believes this to be true, since it has adopted policies to slow the rate of growth in the Mexico City area.[2] Other developing countries have similar policies with respect to their primate cities, and still others likely will begin to move in the same direction.

Slowing the rate of urban growth in the developing countries, or directing it away from primate cities, will not be easy. As noted above, the process is fundamental to economic development, an imperative which limits the options of governments attempting to alter it. Nonetheless, many will try. This study seeks to contribute to that effort by

---

[2] Luis Unikel, Crescencio Ruiz Chiapetto and Gustavo Garcia Villareal, El desarrollo urbano de Mexico (El Colegio de Mexico, Mexico City, 1976) pp. 313-314.

analyzing the urbanization process in Mexico, attempting to identify some of those aspects which offer leverage for changing the pace and direction of the process.

The emphasis in the study is on migration to urban areas. This is an important limitation because natural rates of population growth in urban areas of Mexico, while perhaps less than the 3.5 percent annual rate for the country as a whole, nevertheless are sufficiently high to double urban populations in thirty-five years or less. For example, there is some evidence indicating that the natural rate of increase of the population resident in the Mexico City metorpolitan area in 1960 was 2.0 percent annually between 1960 and 1970, well below the rate for the country as a whole.[3] The population of the Mexico City metropolitan area in 1970 was 8.34 million.[4] Should that population of the area reproduce at an annual rate of 2.0 percent, the population of the area would rise to 14.9 million in 2000, even if net migration between 1970 and 2000 were zero. Accommodating 14.9 million people in the Mexico City area could present enough problems that the Mexican government would consider limiting the natural rate of increase.

Thus, a comprehensive urban policy in Mexico likely will have to deal with natural population increase as well as with migration. However, a

---

[3]Francisco Alba, La poblacion de Mexico: evolucion y dilemas (El Colegio de Mexico, Mexico City, 1977) pp. 89-90, gives data from Ana Marie Goldani, Impacto de la immigracion sobre la poblacion del area metropolitana de la Ciudad de Mexico (El Colegio de Mexico, 1976) which implies the 2.0 percent annual rate.

[4]Unikel and coauthors, El desarrollo urbano de Mexico, p. 26.

policy with respect to natural increase would necessarily require attention to a range of issues, for example family planning, which are outside the scope of this study.

It is worth noting that over periods as long as a decade, births attributable to migrants contribute significantly to measured natural increase in urban areas. In the Mexico City metropolitan area, for example, net migration accounted directly for 36.3 percent of the increase in the area's population between 1960 and 1970. However, children born to migrants in that decade accounted for an additional 33.1 percent of the growth of population.[5] Consequently, migrants, directly and indirectly, were responsible for almost 70 percent of the area's population growth between 1960 and 1970. This suggests that migration should be the focal point of urban policy in Mexico.

Overview of the Research Design

This study seeks to answer two questions: (1) can public policy as applied to rural regions be expected to affect rural-to-urban migration in Mexico, and, if so, what are the more effective instruments for implementing such a policy? (2) can such instruments be applied to alter the distribution of migrants as between Mexico City and other smaller cities?

---

[5]Alba, La poblacion de Mexico, pp. 89-90, citing Goldani, Impacto de la immigracion. Goldani's estimate of 36.3 percent as net migration's share of population growth in the MCVA from 1960 to 1970 is less than Unikel's estimate of 43.2 percent (El desarrollo urbano de Mexico, p. 46). However, the argument that children born to migrants after arrival in the city greatly magnify the contribution of migration to total urban growth is not affected.

The research undertaken attacks the question at two levels. The first is an economic analysis of state-to-city migration in Mexico from 1950 to 1960 and from 1960 to 1970. The analysis makes use of inferential statistics to test hypotheses concerning determinants of gross migration in a cross section of state-city pairs. The tests are structured in the context of multiple equation models of regional labor markets.

A second level of analysis is a more intensive study of urbanization and rural development linkages within the state of Sonora.

The two studies are intended to supplement each other in assembling evidence from an array of nonexperimental data toward an answer of the two policy questions.

The econometric analysis. In theory, public investment in a given region can increase the productivity of labor, thereby shifting labor demand schedules to the right and increasing wages in the region relative to those in other regions. This will attract labor into the more favored region and the ensuing migration flow can impact the wage structure of all other regions as well. The outcome of such a complex series of effects and associated migration patterns can be determined only by general equilibrium analysis.

Models of a multi-sector, multi-regional character often fail to reflect realistic market behavior in treatments of both supply and demand, as, for example, in the case of the input-output model. The addition of mathematical programming models provides some improvement, but all such large-scale models typically posit the existence and form of an entire system of causal relations when, in fact, little is truly known about the structure of large blocks of these systems. Accordingly, we have focused our research effort on estimation of key variables and relationships underlying urban and rural labor market behavior and migration in Mexico rather than on a programming approach which assumes that these variables and relationships are already known.

We have chosen an econometric approach that employs empirial hypothesis testing, including regression and associated techniques. However, the form in which the data are available severely limits the degree of sectoral disaggregation, and the degree to which interregional simultaneity can be treated. The latter limitation is a clear result of the cross-sectional nature of the data: if all regions were included as variables in an inter-regional interaction model, no degrees of freedom would be available for coefficient estimation and hypothesis testing.

The compromise chosen was to structure a highly aggregative regional labor market model with but two sectors, agricultural and urban, and with a limited degree of regional interdependence. Two variants of this model were tested, differing in the treatment of regional interdependence. The first variant considers simultaneous wage and migration determination within urban labor markets, with these effects treated as recursively dependent upon labor market events in rural regions. The second variant treats the likelihood that rural labor market events, in turn, are sensitive to feedback effects from the cities.

Common to both formulations of the model is an evaluation of whether such policy variables as irrigation investment, farm equipment investment, land reform, and state health and education programs can be expected to raise rural productivity. These relationships are specified in a single rural productivity equation that predicts average state rural income per worker.

Both variants of the model are used to ask the question of whether states with higher rural incomes have lower levels of out-migration to cities, taking into account the wage levels of those cities. This relation is specified in a state-to-city migration equation that can reflect the labor supply relation of a state and city pair. It also takes account of, among other variables, the degree of "ruralness" of the state-of-origin, the size of the city at the destination, and characteristics of the at-risk state populations.

Further, in both approaches, the possibility is considered that the urban wage that serves to attract in-migrants is itself affected by that migration. This question is tested via a third equation, an urban labor demand equation that explains a city's wage level, partly in terms of immigration to the city. The consequent simultaneity in the determination of the urban wage and the migration flow requires the application of a simultaneous equations estimation procedure.[6]

The possibility is then considered that urban wages in a given state are affected by rural wages in the rest of that state and by the size of that state's rural labor force. This possibility arises from the role that thriving rural hinterlands have as markets for urban goods and services. The nature of the interdependence implied is that rural public investment programs in a given state may, by increasing rural wages, not only reduce out-migration but also may increase urban wages in the state. An effect

---

[6]For treatment of the simultaneity issue, see Michael Greenwood, "Urban Economic Growth and Migration: Their Interactions," Environment and Planning, vol. 5 (1973), and Richard Muth, "Migration: Chicken or Egg?" Southern Economic Journal, vol. 13 (1971).

may be that cities in the state will increase their attraction for rural migrants. This may have but modest impact upon the state's rural population since the state's rural wages must, of course, have also increased. But it is likely to attract migrants from other states, particularly from those nearby, in which the pace of rural development is slower. To the extent that this effect exists, rural public investment programs can be selectively applied not only to stem rural out-migration, but also to stimulate urban economic development in the states receiving the investments, thus shifting the pattern of migration destinations as among cities. This effect is tested merely by including the average rural income and rural labor force variables in the urban wage equation. A flow diagram that depicts this minimal block recursive structure is shown in diagram 1.

Diagram 1. Block recursive structure

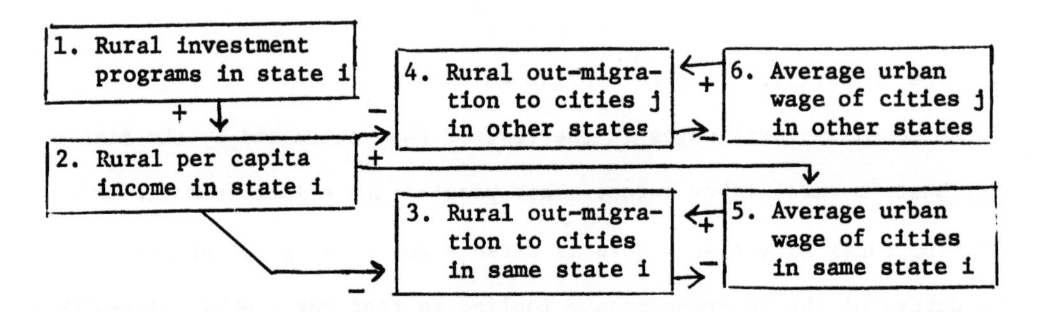

The direction of the arrows shows the direct effect of one variable on another and the signs of the arrows show whether the effect is positive negative.

There are two indirect effects not considered in the first variant of the model as depicted in diagram 1. The first is that the increase in urban wages indirectly induced by the initial rise in rural income will induce a secondary increase in rural income beyond the initial rise. This effect is depicted by the positive arrows running from blocks 3 and 5 to block 2 in diagram 2.

The second indirect effect occurs because the growth of urban economic activity indirectly induced by the initial investment in agriculture will facilitate the flow of technical knowledge and inputs to agriculture, thus stimulating the growth of agricultural productivity and wages in addition to the stimulus given by the initial investment. This effect is depicted in diagram 2 by the positive arrow running from block 5 to block 2 and by the two positive arrows linking blocks 6 and 7. For purposes of this analysis the role of urban development in stimulating higher agricultural productivity is reflected in higher urban wages of blocks 5 and 6.

These possibilities require that all parameters of the model be simultaneously estimated. They require treatment not only of urban-rural interdependence within states receiving rural public programs, but also of urban-rural interdependence in states receiving in-migrants. That is, migrants reaching destination cities may lower urban wages and consequently cause incomes in the surrounding rural hinterlands to fall. Such wage effects may further affect migration patterns. All these possible relationships are tested by treating rural income, the urban wage, and state-to-city migration as simultaneously determined in a system of three equations. This simultaneous structure is shown in diagram 2.

Both variants of the regional labor market model are tested in the course of the econometric study. Both contain three equations: a rural income equation, an urban wage equation, and a state-to-city gross migration equation. They differ in their treatment of regional interdependence.

Diagram 2. Simultaneous structure

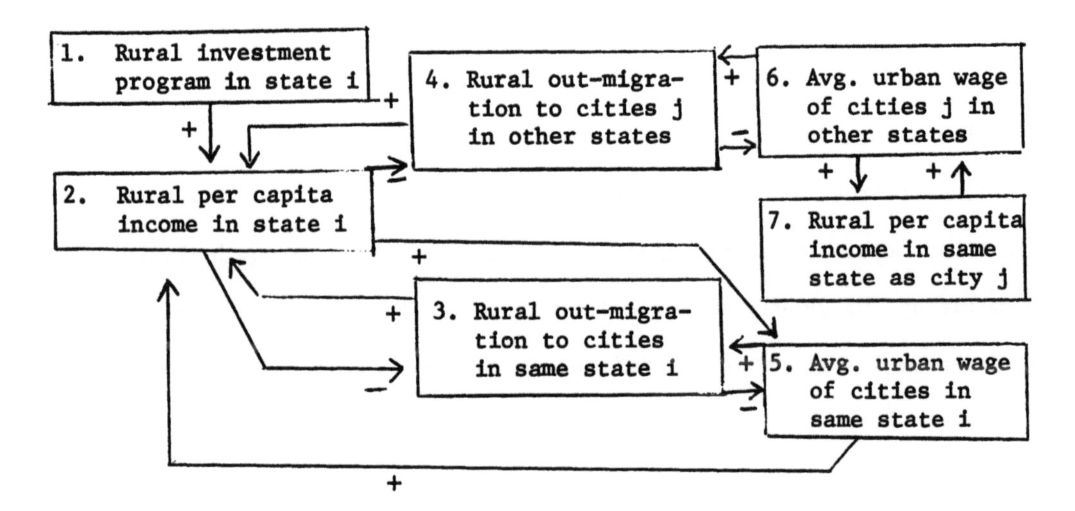

<u>The case study for the state of Sonora</u>. The results of the regression analyses suggest that policies to increase agricultural income both reduce the amount of migration from rural areas and stimulate migration to the urban centers serving those areas. The results suggest, therefore, that the answer to the two policy questions posed on page 6 is "yes," and that policies to stimulate agricultural income have some promise for implementing an urbanization policy in Mexico.[7]

The regression analyses, however, are based on highly aggregated data and the possibility that they are consequently subject to bias cannot be avoided. We pay close attention to this problem and believe that it is not so severe as to invalidate the analyses or the conclusions we draw

---

[7]The subsequent analysis supports this statement. However, the analysis also shows that an agricultural income policy alone would not likely be sufficient to implement a national urbanization policy. Other measures would be needed. This is discussed further in the final chapter of this report.

from them. We think it useful, nonetheless, to supplement our econometric work with a study of the process of agricultural development and attendant urban growth in the state of Sonora. The study of Sonora does not deal explicitly with migration. Its focus instead is on the relationship between the expansion of agricultural production and productivity in a region and the growth of urban centers in the region oriented to serve agriculture. The study thus is relevant to the hypothesis explored in the econometric analysis that investments in agriculture not only may slow the rate of rural-to-urban migration, they may also influence the direction of migration by increasing income and employment in urban areas indirectly favored by the investments. If the results of the study are consistent with those of the econometric analysis, and they are, the credibility of the latter is strengthened. The case of Sonora of course is not typical of all of Mexico. No single state experience is. Consequently we do not claim that the analysis of Sonora confirms the results of the econometric analysis in all detail. However, the experience of Sonora, with respect to the two principal questions addressed in this study, is thoroughly consistent with what one would expect from the econometric analysis.

Layout of the Study

Chapter 2 presents the principal quantitative dimensions of urban population growth in Mexico between 1940 and 1970, focusing particularly on migration and characteristics of migrants relevant to understanding their behavior. While most of the discussion deals with Mexico, experience in other countries thought to be relevant to Mexico also is considered.

The structure and results of the econometric analysis are presented in chapters 3 through 6. In chapter 3, the rural productivity equation is evaluated. Variables that are sensitive to policy as well as those that condition rural productivity are tested through multiple regression analysis. The results of both single equation and simultaneous equations formulations are shown.

Chapter 4 presents the structure and findings of the migration analysis. Several formulations of the economic rate of return hypothesis to explain migration are tested. Hypotheses relating to the role of labor market information in decisions to migrate, uncertainty, and personal attributes of the at-risk population are also tested. All tests make use of a simultaneous equations estimation procedure; for most tests, the sample is split to separate migrants from the more rural states from those from the more urbanized states.

The determinants of the urban wage, including the effects of rural hinterland markets, are analyzed in chapter 5. A simultaneous equations estimation procedure is used for estimating the coefficients in this relationship and testing hypotheses about it.

Chapter 6 discusses the principal implications of the analysis for urbanization policy in Mexico. The equations of chapter 3, 4, and 5 are used to give a rough quantitative indication of the effect of a rural development program in reducing migration to Mexico City from rural areas.

Chapter 7 presents the case study of Sonora, and Chapter 8 summarizes our findings and policy conclusions.

Chapter 2

ASPECTS OF URBAN POPULATION GROWTH IN MEXICO

Dimensions of Urban Population Growth

The rapid growth of urban population in Mexico, particularly in the three decades between 1940 and 1970, is evident in table 1. By 1970 Mexico was one of the most urbanized of the developing countries, especially among the larger ones. Its rate of urban population growth between 1940 and 1970 was certainly among the highest in the world, although comparisons with other countries are difficult because of differing census definitions of urban places. Among the larger Latin American countries only Venezuela and Peru had faster rates of urban population growth than Mexico.[8]

Urban development in Mexico, as in many of the less-developed countries, has been dominated by a single area: in this case, Mexico City. In 1900, the Mexico City urban area (MCUA), with 345,000 people, contained 24 percent of the country's urban population.[9] By 1960, the proportion had increased to 38.5 percent (4.91 million). It then declined slightly to 38.0 percent in 1970 (8.36 million).[10] The relative importance of Mexico City in the hierarchy of cities can be measured by the so-called primacy index. This is the ratio of population in the MCUA to the sum of

---

[8] Luis Unikel, Crescencio Ruiz Chiapetto and Gustavo Garcia Villareal, El desarrollo urbano de Mexico (El colegio de Mexico, Mexico City, 1976) p. 60.

[9] We adopt Unikel's definition of the MCUA: most of the Federal District plus, since 1940, adjacent parts of the state of Mexico.

[10] Unikel, El desarrollo urbana de Mexico, p. 27.

Table 1. Population Growth in Mexico, By Decade, 1900-1970

Year	Population (thousand persons)		Average annual per-cent growth from previous census year		Urban population as per cent of total
	Total	Urban[a]	Total	Urban	
1900	13,607	1,434	--	--	10.5%
1910	15,160	1,783	1.1%	2.2%	11.8
1921	14,335	2,100	-0.5	1.5	14.6
1930	16,553	2,891	1.7	3.9	17.5
1940	19,649	3,928	1.5	3.2	20.0
1950	25,779	7,210	2.7	6.1	28.0
1960	34,923	12,747	3.1	5.9	36.5
1970	49.050	22,004	3.5	5.6	44.9

Source: Luis Unikel, El Desarrollo Urbano de Mexico, p. 27.

[a]Urban population is defined as persons living in places with 15,000 or more inhabitants. This is the figure used by Unikel to distinguish urban from non-urban places in Mexico. The figure is based on a detailed analysis of various sizes of communities with respect to social and economic characteristics generally associated with urban life. Unikel found that for most of these characteristics the sharpest distinctions were between places with more and less than 15,000 people.

populations in the 2nd, 3rd,...nth largest cities in the country. Table 2 shows the primacy indexes for selected years since 1970. The persistent dominance of the MCUA is evident. While it lost a bit of ground to the second largest city, Guadalajara, it still was six times as large in 1970. With respect to the other groups of cities shown, the MCUA grew faster between 1940 and 1970.

It is apparent that Mexico City continues to dominate the country's urban landscape. Nonetheless, the primacy indexes show that significant growth has occurred also in regional urban centers. Table 3 provides additional information in this respect. From 1940 to 1950 the share of the MCUA in total urban population rose, primarily at the expense of Guadalajara and "all other" category. The share of the MCUA declined again from 1960 to 1970 as did the shares of Monterrey and Guadalajara. Those of the border cities and the agriculturally based cities increased slightly, while the share of "all other" areas increased, reversing the trend of the two previous decades.

If it is thought desirable that the concentration of population in the MCUA, Guadalajara, and Monterrey should diminish, then the declining shares of these cities from 1960 to 1970 are encouraging. The increase in the share of the "all other" category of cities is particularly noteworthy. Cities with from 15,000 to 99,999 inhabitants in 1970 (other than those cities identified on table 2) were responsible for this increase. This set of cities had 19.4 percent of the nation's urban population in 1960 and 22.4 percent in 1970. There were 143 of these cities in 1970 spread widely throughout the country. The relative growth of these places from 1960 to 1970 suggests a vitality in small- and medium-sized

Table 2. Primacy Indexes for Mexico

Number of cities compared	Year			
	1940	1950	1960	1970
2 cities	6.5	7.2	6.1	6.1
4 cities	2.7	2.9	2.7	2.8
6 cities	2.0	2.2	2.1	2.2
8 cities	1.6	1.8	1.8	1.9
10 cities	1.4	1.6	1.6	1.7

Source: Unikel, El desarrollo urbano de Mexico, p. 57. The primacy index is as follows:

$$\frac{P_1}{P_2 + P_3 + \text{---} + P_n}$$

where P_1 is the population in the largest city, P_2 is that of the second largest city, and so on. The two city index, therefore, is the ratio of P_1 to P_2, the 4 city index is the ratio of P_1 to the sum of P_2, P_3 and P_4, etc.

Table 3. Shares of Selected Cities or Urban Areas in Mexico's Urban Population

City or urban area	Year 1940	1950	1960	1970
		(percent)		
Mexico City urban area	39.7	41.2	38.5	38.0
Guadalajara	6.1	5.6	6.7	6.2
Monterrey	4.8	4.9	5.5	5.0
6 border cities[a]	3.5	5.4	6.6	6.7
3 agriculturally based cities in the northwest[b]	1.4	1.7	2.0	2.1
all other urban areas	44.5	41.2	40.7	42.0
Total	100.0	100.0	100.0	100.0

Source: Unikel, El desarrollo urbano de Mexico, pp. 158-159

[a]Matamoros, Nuevo Laredo, Reynosa, Cuidad Juarez (Texas border), Mexicali and Tijuana (California border).

[b]Culiacan (state of Sinaloa), Cuidad Obregón and Hermosillo (state of Sonora).

towns that bodes well for a policy to decentralize the urban population. Indeed, the performance of these towns in the 1960s indicates that decentralization already was under way.

Sources of Urban Population Growth

Urban areas grow through natural increase, by net migration, and by extending their boundaries to incorporate neighboring communities. In addition, the nationwide figures for urban population will reflect the shift of centers from nonurban to urban status as their populations reach 15,000. This latter process accounted for 19.2 percent of measured urban growth in Mexico between 1940 and 1950, for 13.8 percent between 1950 and 1960, and for 4.2 percent between 1960 and 1970. For the period 1940 to 1970, reclassification accounted for 10 percent of measured urban growth nationwide. The incorporation of neighboring nonurban communities (mostly in the urban area of Mexico City and to a lesser extent those of Guadalajara, Monterrey, and Puebla) was responsible for 12.7 percent of urban growth between 1940 and 1950, and for 8.3 percent of that between 1950 and 1960.[11]

Of course, urban "growth" through reclassification and incorporation of neighboring communities is not wholly additive to growth from natural increase and net migration. Towns which increase in size to have at least 15,000 inhabitants do so by either or both of these processes, and the incorporated neighboring communities may grow for the same reasons. Unikel presents estimates showing that natural growth and net rural-to-

---

[11]All of these figures are from Unikel, El desarrollo urbano de Mexico, pp. 54-55. Unikel does not give a figure showing how much incorporation of nonurban commmunities contribute between 1960 and 1970.

urban migration accounted for 86 percent of the increment of urban population in 1940-1950, 88 percent of the increment between 1950 and 1960, and 91 percent of the increment between 1960 and 1970.[12]

Natural increase and net migration. It is clear that much of the greater part of the urban growth in Mexico since 1940 was attributable to natural increase and net migration. As indicated in chapter 1, we are particularly interested in the contribution of migration to urban growth. Table 4 shows this by decade for the period 1940-1970 for the three largest cities and for all other urban areas as a group. Several features of the table are noteworthy. One is the decline over each decade in the relative contribution of migration to total growth, the only exception being Mexico City in 1960-70 compared with 1950-60. For the country as a whole, such a decline would be expected unless natural rates of increase in nonurban areas were much higher than in urban areas. If these rates are about the same in both areas, then migration will increase urban population relative to nonurban population, thus increasing the urban base for natural increase relative to the nonurban base for migration. Under these circumstances, migration would have to increase at an increasing rate to maintain its share of total urban increase.

Unikel's data shows that although the natural rate of increase in nonurban areas was higher than that in urban areas throughout the period 1940-1970, the differences diminished sharply in those years.[13] Hence

---

[12]Unikel, El desarrollo urbano de Mexico, relation between table 1-3, pp. 44-46

[13]Unikel, El desarrollo urbano de Mexico, p. 47.

Table 4. Contribution of Net Migration to Population Growth in Urban Areas, 1940-1970[a]

(population figures in thousands)

	Mexico City	Guada-lajara	Mont-errey	All Other	Total Urban
1940-50					
Increase in urban population					
Total	1228.6	158.6	161.9	1272.1	2822.2
Migration	847.2	97.6	97.0	613.5	1655.3
Migration					
% of total increase	68.9	61.6	60.0	48.2	58.7
% of total migration	51.2	5.9	5.9	37.0	100.0
1950-60					
Increase in urban population					
Total	1930.9	395.4	336.5	2220.3	4883.1
Migration	739.1	227.9	172.3	621.6	1760.9
Migration					
% of total increase	38.3	57.6	51.2	28.0	36.1
% of total migration	42.0	12.9	9.8	35.7	100.0
1960-70					
Increase in urban population					
Total	3445.1	570.2	396.4	4021.5	8433.2
Migration	1488.5	215.9	65.2	979.9	2749.5
Migration					
% of total increase	43.2	37.8	16.5	24.4	33.8
% of total migration	54.1	7.9	2.4	35.7	100.0

Source: Unikel, El desarrollo urbano de Mexico, pp. 44-46.

[a]The estimates exclude increases due to expansion of urban areas to incorporate previously rural areas and to changes in classification from rural to urban by growth from less than to more than 15,000 inhabitants. Hence the part of total increase not due to migration is due to natural increase. As noted in the text, natural increase and migration accounted for about 90 percent of total urban growth between 1940 and 1970.

both the population base in urban areas and the rate of growth of the base rose relative to nonurban areas. The fall in the share of migration in total urban population growth, therefore, is not surprising.

This reasoning applies with less force to any given urban area because the potential pool of migrants from which it can draw is much larger relative to its own population than is true for urban areas as a whole. Nonetheless, in each of the three decades the contribution of migration to total growth fell for all the cities and groups of cities listed in table 4 except, as noted, in Mexico City from 1960-70. The increase in the contribution of migration to Mexico City's growth from 1960 to 1970 is impressive evidence of the vitality and attractiveness of that area as a place to live and work, relative to other places, despite Mexico City's well-advertised problems of congestion and pollution. Whether migration to the area continued at the same pace in the 1970s will not be known until data are available from the 1980 census. Note also that Mexico City's share of total migration to urban areas increased sharply in 1960-70 relative to 1950-60 and at 54.1 percent was even higher than in 1940-50.

In each area the proportion of growth attributable to migration would be greater if the children born to migrants were counted as part of the contribution of migration to growth. The only estimate of this indirect effect we have seen is that for Mexico City, cited above, indicating that the children born to migrants contributed almost as much as the migrants themselves to growth of that area between 1960 and 1970. If this estimate is reasonably representative of the indirect contribution of migration to growth of population in other urban areas, then the figures in table 4 would rather considerably understate the share of migration in total urban growth.

Some Characteristics of Urban Migrants

Most economists and many sociologists seek to explain migration as a response to relative differences in economic opportunity between places of origin and destination. The analysis of migration in this study is based fundamentally on that hypothesis, although considerable attention is given also to the influence of noneconomic factors on decisions to migrate. The following three chapters present a variety of materials relevant to the statistical analysis of migration. The remainder of this chapter is devoted to a discussion of some of the characteristics of migrants which appear relevant to the explanation of their behavior. This "profile" of the migrant will help to put flesh on, if not impart spirit to, the statistical bones analyzed later.

We are interested in knowing more about the migrant than the data on migratory flows and economic conditions alone can tell us: how he compares with his fellows in places of origin and destination; how he fares after arrival in the city; how he views the migratory experience, and his aspirations. While the discussion deals primarily with materials on migrants in Mexico, some attention is given also the literature on migrants in other countries where this contributes to better understanding of the Mexican situation.[14]

---

[14]Unless otherwise noted, the following discussion deals with all migrants, not just those from rural areas.

Between 1930 and 1960 migrants in Mexico were selective with respect to age and sex.[15] This is evident in table 5. In each of the three decades considered, the percentage of both male and female migrants in ages 15-29 was substantially greater than the percentage representation of these ages in the total population. The percentages of migrants in both younger and older age groups is correspondingly less than the share of these ages in the total population. Female migrants outnumbered males by a considerable margin in all three decades, although the relative difference declined. The selectivity by sex was especially marked in the age groups 10-14 and 15-24.

The age-sex distribution of migrants in Mexico is similar to that found in other countries in Latin America. In a study of migration to Greater Santiago, Herrick found that 63 percent of migrants sampled had arrived before age 26. Women greatly outnumbered men, there being 150 females for each 100 males in all age groups over 14 years. In the group ages 15-24 there were 194 women for each 100 men.[16] A similar pattern was found by Germani in a study of migrants to Greater Buenos Aires and by Flinn in a study of rural-to-urban migration in Colombia.[17]

---

[15] Comparable data for 1960-1970 are not available at this writing.

[16] Bruce Herrick, Urban Migration and Economic Development in Chile (Cambridge, M.I.T. Press, 1965) p. 108.

[17] Gino Germani, "Inquiry Into the Social Effects of Urbanization in a Working Class Sector of Greater Buenos Aires," in Philip M. Hauser (ed.) Urbanization in Latin America (Paris, UNESCO, 1961); William L. Flinn, "Rural to Urban Migration: A Colombia Case," Research Paper, no. 19 (Land Tenure Center, University of Wisconsin, Madison, July 1966).

Table 5. Age and Sex Distribution of Net Interstate Migration and of Total Population in Mexico, and Male-Female Ratio of Migrants

Sex and Age at end of period	Percent distribution of net migration			Percent distribution of total population				Male-Female ratio of migrants (x 100)		
	1930-40	1940-50	1950-60	1930-40	1940-50	1950-60		1930-40	1940-50	1950-60
Men										
10-14	16.6%	11.7%	12.8%	16.7%	16.3%	17.4%		82.1	80.2	81.6
15-24	32.3	35.3	39.2	25.1	27.2	27.2		82.3	84.7	88.2
25-29	14.4	15.6	13.8	10.3	10.6	10.1		99.5	116.1	105.5
30-39	15.2	16.3	14.7	18.3	16.0	16.3		83.0	103.9	102.1
40+	21.5	21.1	19.5	29.6	29.4	29.0		84.6	85.6	92.8
Total	100.0%	100.0%	100.0%	100.0%	100.0%	100.0%	Average	87.5	90.9	92.0
Women										
10-14	17.6	13.2	14.4	15.4	15.1	16.5				
15-24	34.7	38.0	40.9	24.1	26.9	26.6				
25-29	12.7	12.2	12.0	11.1	10.8	10.6				
30-39	13.1	14.3	13.4	19.2	16.5	17.1				
40+	22.2	22.3	19.3	30.2	30.7	29.2				
Total	100.0%	100.0%	100.0%	100.0%	100.0%	100.0%				

Source: Dinamica de la poblacion de Mexico (El Colegio de Mexico, Mexico City, 1970) Based on table IV-4.

The usual explanation of the age selectivity of migrants combines several elements. One is that the young generally combine fewer responsibilities with a greater willingness to accept risks. In addition, the young have more working years remaining than the older population; hence the present value of the income stream the young can expect to capture from a move likely will be greater than that the older population can expect. For these reasons, the young are more mobile than older persons.

The selectivity by sex is not so readily explained, but the conventional wisdom holds that the disadvantage of women relative to men is greater in the competition for agricultural jobs than for urban jobs. The rapid pace of urbanization all over Latin America has greatly expanded the opportunities for domestic service employment in particular. These positions typically are held by women.

There is some evidence that male migrants in Mexico are better educated and prior to migration held higher paying jobs than males generally in the places of origin. This evidence, admittedly quite slender, is given by Browning and Feindt, in a study of male migrants to Monterrey.[18] The study is based on a sample survey taken in 1965. Three cohorts of migrants were identified: those who arrived in Monterrey before 1941, those who arrived between 1941 and 1950, and those who arrived between 1951 and 1960. For each cohort the percentage having completed six years of primary school and with nonagricultural job experience prior to

---

[18] H. L. Browning and Waltrant Feindt, "Selectivity of Migrants to a Metropolis in a Developing Country: A Mexican Case Study," Demography, vol. 6, no. 4, November 1969.

migration were calculated and compared with the percentages for all males over 15 years old in the areas of origin.[19] In general the percentages for migrants were higher, indicating that migrant males were better educated and had more nonagricultural job experience than males in areas of origin. The only exceptions were in comparisons with Zone One, that containing the Federal District. Forty-eight percent of the males in that zone in 1940 had completed six years of primary education while only 39 percent of male migrants from there to Monterrey before 1941 had done so. This situation was subsequently reversed, however, some 62 percent of the Zone One migrants arriving between 1951 and 1960 having completed six years of primary school compared with 51 percent of all males in the zone in 1960. The first zone was also the only one in which the percentage of males with nonagricultural job experience was higher than the percentage for migrants.

It is also of interest to compare migrants with natives in urban areas of destination with respect to job experience and education. Tables 6 and 7 permit such comparisons for Mexico City and Monterrey. In neither table were the native-migrant differences tested for statistical significance. The native-migrant differences by occupation were apparently more marked in Monterrey, the migrant population containing a rather considerably greater proportion of unskilled manual workers than the native population and rather fewer white collar, professional, and managerial workers. Balan, the source of these data, suggests that the smaller occupational differences in Mexico City may reflect the fact that a greater proportion

---

[19]Five such areas were identified, ranging from highly urban (the Federal District and four other zones) to higher rural (the state of Zacatecas and two others).

Table 6. Mexico City: Distribution of Economically Active Males According to Occupation, By Migratory Status and Age, 1960

Age and Status		Occupation					
		Professional	Office workers	Sales	Production workers	Nonproduction workers	Service
12-29	Native	6.3%	20.4%	10.6%	47.8%	10.2%	4.7%
	Migrant	5.8	17.4	10.6	44.8	11.1	10.6
30-44	Native	7.2	18.1	15.8	48.8	5.8	6.1
	Migrant	6.7	15.9	15.9	43.4	10.1	7.9
45-60	Native	8.0	19.1	16.7	42.3	8.8	5.0
	Migrant	7.7	17.0	18.7	39.9	8.5	8.2
60+	Native	8.6	15.7	24.9	33.5	9.6	7.6
	Migrant	8.7	20.6	24.5	27.3	8.7	10.2

Source: Jorge Balan, "Migrant-Native Socioeconomic Differences in Latin American Cities: A Structural Analysis," Latin American Research Review, vol. IV, no. 1, Spring 1969, p. 19. Based on a 1.5 percent Population Census sample of the Federal District.

Table 7. Monterrey: Socioeconomic Differences Between Native and Migrant Males, 21 to 60, 1965

Socioeconomic category	Natives	Migrants (by size of place of birth)			
		Total	Up to 5,000	5,000 to 20,000	20,000 & more
Education					
Less than primary school completed	30.2%	58.7%	67.1%	52.1%	38.0%
Primary school completed	30.1	30.6	17.4	25.3	25.6
Secondary education	26.8	14.7	11.3	16.9	23.7
University education	12.9	6.0	4.2	5.8	12.7
Total	100.0%	100.0%	100.0%	100.0%	100.0%
Occupation					
Unskilled manual	28.4	37.2	42.9	31.7	24.2
Skilled manual	40.5	43.8	40.9	50.0	46.8
White collar	21.2	13.4	12.2	12.1	19.3
Professional, managerial	9.9	5.5	4.0	6.2	9.6
Total	100.0%	99.9%	100.0%	100.0%	99.9%

Source: Balan, "Migrant-Native Socio-economic Differences," p. 20. Based on the same survey of Monterrey as employed by Browning and Feindt, "Selectivity of Migrants."

of migrants there originated in other urban areas. While precise information is not available on the origins of migrants to Mexico City, it is known that the city has long held an unmatched attraction for ambitious persons seeking to rise in business, the professions, and the bureaucracy. Such persons are likely to come from regional urban centers and to already possess, or be able quickly to acquire, the educational and other skills necessary to achieve their occupational goals. Migrants to Monterrey, on the other hand, are more likely to come from rural areas and to find it more difficult, therefore, to compete with natives for higher paying and more demanding urban jobs.[20]

Migrants to Monterrey were clearly less well educated than natives. This was true regardless of size of place of origin, although the educational disadvantages of migrants from larger places was substantially less than that of those from smaller places. Migrants apparently occupied middle ground with respect to education: they had more than the average in their places of origin,[21] but less than the average for natives of Monterrey.

Table 7 indicates that the native-migrant differences with respect to education are greater than those for occupation. This is particularly true for migrants from places of less than 5,000 inhabitants, i.e., those most likely to be considered "rural." At first blush this pattern is somewhat

---

[20] Balan, "Migrant-Native Socio-economic Differences," p. 19. In the sample of migrants on which table 7 is based, 62 percent were born in localities of less than 5,000 inhabitants. However, only 42 percent were living in communities this size just before migrating to Monterrey.

[21] See discussion above of Browning-Feindt study.

surprising. One expects to find strong positive correlation between place on the educational ladder and place on the occupational ladder. There is such a correlation in table 7, but it seems weaker than might be expected. It may be that the real correlation is stronger but that it is obscured by the grossness of the occupational categories.

It is of interest that Browning found a similar situation in a study of native-migrant differences in Mexico City. His results are summarized in the following quotation.[22]

> Migrants do not differ greatly in most respects from the native population in some socioeconomic characteristics. The major exception is education, in which the migrants, especially workers, are much inferior to natives. This does not seem to have much effect, surprisingly, when it comes to occupational and income differences. Natives are superior, but the difference is not large. And when the standard of living indicators are taken from the census data (availability of running water in households, sewage facilities, separate bathrooms, ownership of radio and T.V.) there is virtually no difference between the two groups.

Studies of native-migrant differences in other cities of Latin America suggest that the pattern observed in Mexico City and Monterrey may be somewhat unusual: there other studies show small native-migrant differences for education as well as occupation.[23]

---

[22]Harley L. Browning, "Urbanization and Modernization in Latin America: The Demographic Perspective" in Glenn H. Beyer (ed.), The Urban Explosion in Latin America (Ithaca: Cornell University Press, 1967) p. 91.

[23]Balan, "Migrant-Native Socio-economic Difference," p. 17, presents data on education of migrants and natives in Greater Santiago taken from a labor force survey conducted by the Instituto de Economia, University of Chile in June 1958. Herrick, Urban Migration, pp. 80, 86, gives occupational data for migrants and natives in Greater Santiago in 1963. Balan also provides native-migrant comparisons by education and occupation for San Salvador and Guatemala City. In all the instances cited, the native-migrant differences for both education and occupation are small.

It would be very useful to know something about the migrant's unemployment experience after arrival in the city, particularly as it compares with that of urban natives.[24] We know of no such information for Mexico. In fact, to our knowledge the only such information for any country in Latin America is that provided by Herrick in his study of migration in Chile. Since the occupational status of migrants is much like that of urban natives in both Mexico and Chile, Herrick's data on native-migrant unemployment experience in Greater Santiago may be of interest. Of course, no claim is made that the unemployment experience of the two groups is necessarily the same in Mexico as in Chile. Herrick's data are given in table 8. They leave little doubt the migrants to Greater Santiago fared better than natives in finding jobs. This was true for both men and women and for all but one of the age groups considered.

As indicated above, we have no information on the comparative unemployment experience of urban natives and migrants in other Latin American cities. There is some information, however, on the experience of migrants alone in a number of cities. Germani, in the study of Buenos Aires cited earlier, states that only 2-3 percent of the migrants were unemployed at the time of the survey. Recent migrants fared less well, however. Only 50 percent of them reported they had worked full time in the previous twelve months; one-third said they had worked only six months or less. In a study of migrants to Lima, Matos Mar reported that only one percent

---

[24] It would be even more useful to have information about the migrant's unemployment experience before departure as well as after his arrival. The predeparture information, however, is not available.

Table 8. Unemployment Rates of Natives and Migrants In Greater Santiago, 1963

Sex and Age	Natives	Migrants
Men	7.2%	4.6%
Women	4.9	3.1
Ages		
15-19	14.0	8.8
20-34	6.4	4.8
35-44	3.5	3.6
45-64	4.9	2.4
65+	6.4	4.0

Source: Bruce Herrick, Urban Migration and Economic Development in Chile (Cambridge: M.I.T. Press, 1965).

of those interviewed said they were unemployed. Seventy-one percent reported they had "stable" employment and 27 percent were "casually" employed.[25]

As far as they go, these scattered data suggest that migrants to Latin American cities suffer little permanent unemployment and that they may well do better than urban natives in this respect. This generalization possibly is not as applicable to migrants from strictly rural areas as it is to the whole population of migrants, since those from rural areas generally are less well educated and probably have less nonagricultural job experience when they arrive in the cities. Nonetheless, the information available seems to justify Nelson's conclusion that although the evidence on migrant employment experience is fragmentary, "it seems to indicate consistently for a number of cities that most migrants who seek employment find it reasonably quickly."[26] This finding is particularly significant when it is related to information about differences between rural and urban incomes. In Mexico in the 1960s, urban wages in service industries were 50 percent to 100 percent greater than average labor

---

[25]Germani, "Inquiry into the Social Effects of Urbanization," p. 220; Jose Matos Mar, "Migration and Urbanization--the Barridas of Lima: An Example of Integration into Urban Life,: in Hauser (ed.), Urbanization in Latin America, p. 180.

[26]Joan M. Nelson, Migrants, Urban Poverty and Instability in Developing Nations Harvard University Center for International Affairs, Occasional Paper no. 22, September 1969 p. 18. Nelson's conclusion includes, but is not limited to, cities in Latin America.

income in agriculture.[27] The difference between industrial wages and agricultural income probably was even greater. The wage differences would be a powerful spur to migration from rural to urban areas, if the probability of getting an urban job were high. The urban employment experience just recounted suggests that this probability in fact was high.

The combination of large urban-rural income differences and high probability of finding urban employment suggests that devising effective policies to reduce rural-to-urban migration may prove very difficult. Such policies would have to overcome, or at least weaken, powerful market incentives to migrate. We consider some such policies in this study and find that they can reduce migration to cities. However, the reductions are relatively small, leading us to caution against the notion that rural-to-urban migratin in Mexico (and by inference in other developing countries) can be readily slowed by policies that nonauthoritarian governments would be willing to adopt.

This survey of socioeconomic characteristics of migrants and their experiences in urban areas yields a rather definite image of the typical migrant in Mexico, and other Latin American countries. The migrant is more likely to be female than male and more likely to be of young working age than old. The migrant is clearly better educated than the people who stay behind and, in countries other than Mexico, not much inferior in

---

[27]See table 15, chapter 4, the differences between CWAGE60 (annual city wage in 1960) and AGINC60 (average annual agricultural income in 1960) for rural and urban states. These are averages. The differences between particularly low agricultural income in some states and particularly high wages in some cities are much greater than 100 percent.

this respect to natives of the cities to which the migrant moves. The migrant in Mexico, despite clear educational disadvantages compared with urban natives, occupies only a marginally inferior position in the urban occupational hierarchy. This is true also of migrants to other cities in Latin America. While there are no data concerning urban unemployment experience of the Mexican migrant, those in other Latin American cities seem to have fared reasonably well in this respect. In Mexico, migrants who leave agriculture for an urban job can expect a substantial increase in income.

While the information is quite fragmentary, we now know something about the age, sex, educational, occupational, and income characteristics of migrants in Mexico, and how they differ in these respects from rural populations of origin and the urban populations of destination. This information will be useful in formulating hypotheses to analyze migration to urban areas in Mexico in the years between 1950 and 1970, and in interpreting the results, the task of the next three chapters.

Chapter 3

DETERMINANTS OF AGRICULTURAL INCOME

Public investment in rural regions is one of the policies which may have an effect in reducing rural-to-urban migration if the policy results in a significant increase in rural incomes. If the policy should also increase the income of smaller cities in rural regions, then the result would be not only a decline in rural-to-urban migration but also a diversion of some of it from larger to smaller cities.

Agriculture, directly or indirectly, is the basis for most rural income. Consequently, in this chapter the relation between rural public investment and agricultural income will be examined.

Specification

The production of agricultural goods (a) may be characterized with a production function in which current agricultural output Y_{at} increases with increases in current labor inputs L_{at}, capital K_{at}, public infrastructure $\mathcal{K}_{at}$, land A_{at}, and technology T_{at}

$$Y_{at} = F(L_{at}, K_{at}, \mathcal{K}_{at}, A_{at}, T_{at}). \quad (1)$$

Further, technology reflects the current level of knowledge H_{at}, and the institutionalized form of productive organization I_{at}. More knowledge raises the level of technology, but the manner in which institutional forms affect technology depends upon how the norms and incentives embodied in the institutional structure impact behavior toward innovation and deployment of knowledge

$$T_{at} = g(H_{at}, I_{at}), \qquad (2)$$

so that equation (1) can be written as

$$Y_{at} = F[L_{at}, K_{at}, \mathbb{K}_{at}, A_{at}, g(H_{at}, I_{at})]. \qquad (1')$$

When average agricultural income per worker is defined as agricultural output per worker, equation (1') becomes

$$(Y/L)_{at} = L_{at}^{-1} F[L_{at}, K_{at}, \mathbb{K}_{at}, A_{at}, g(H_{at}, I_{at})]. \qquad (3)$$ [28]

If the production function is of the Cobb-Douglas form, equation (3) can be written

$$(Y/L)_{at} = \alpha L_{at}^{b_1 - 1} K_{at}^{b_2} \mathbb{K}_{at}^{b_3} A_{at}^{b_4} H_{at}^{c_1} I_{at}^{c_2} \qquad (4)$$

and, with a constant returns to scale assumption,

$\sum_h b_h = 1$, $b_h > 0$ both for all h, and the c's are scalar terms. (5)

Definition and Measurement of Variables

Measures for the specified variables were obtained for thirty-one states in Mexico from agricultural and population censuses for 1950 and 1960, as well as from other sources.[29] Complete sets of data for the capital,

[28] Assumes that the share of labor in agricultural output is the same for all producing units. It is not, but our results indicate that the differences are not so great as to invalidate the assumption.

[29] A description of the data sources is provided in the appendix.

technology, and institutional variables were not available; consequently, these variables are represented by proxies assumed to represent the influence of the classes of variables for which complete information is not available. Of course this is an imperfect arrangement. Yet, it yields useful insights into the determinants of agricultural income in Mexico and how they may be influenced by public policy.

Estimation of the production function specified in equation (4) likely would be biased by collinearity among some of the independent variables. Arable land, for example, is highly correlated with agricultural labor force (r - .86).

We took two measures to reduce the collinearity problem. One was to eliminate arable land from the regression equation. We could have as logically excluded labor and included land. However, in the subsequent analysis of migration and of interactions between urban and rural labor markets, we obviously need a labor variable.

The second measure to reduce the collinearity problem was to rewrite several of the variables of equation (4) in ratio form, as follows:

$$(Y/L)_{at} = \alpha\, L_{at}^{b_1 + b_3 - 1} \quad (K/A)_{at}^{b_2} \quad (K/L)_{at}^{b_3}\ H_{at}^{c_1}\ I_{at}^{c_2} \qquad (4')$$

This is the form in which the data have been used in the estimation of the agricultural production function. To reflect the Cobb-Douglas form of the specification, the initial estimation is in the logarithms of all variables. Subsequently, however, a linear estimation also is obtained. Following is a list of the measured variables and their acronyms.

Dependent variable

AGINC(t): state agricultural output per worker per year in pesos, 1950 and 1960.

Independent variables: factor inputs

RURLF(t): state rural labor force, 1950 and 1960, a proxy for L in equation (4'). RURLF differs from L by the amount of unemployment or underemployment in agriculture, for which there are no reliable data.

TRCPH(t): number of tractors per 1000 hectares of state arable land, 1950 and 1960, a proxy for K/A in equation (4').

IRRPL(t): number of hectares of state irrigated land per member of the rural labor force, 1960, a proxy for K/L in equation (4').

Independent variables; knowledge and institutions

EDTPP(t): state primary school personnel per thousand population, 1949 and 1959.

PPLIT(t): percent of state population aged six or more literate, 1950 and 1960.

EDTPP and PPLIT are proxies for H in equation (4').

PABEJ(t): percent of state arable land in *ejidos*, a proxy for I in equation (4').[30]

---

[30] An *ejido* is a group of farming units organized under the Mexican land reform program. In some *ejidos*, the land is farmed as single communal units, but in most, each farmer manages his plot individually. *Ejido* lands are property of the national government and, under the law, may not be sold or rented by the farm operators, In fact, rental is a common practice.

Means, Coefficients of Variation, and Correlation Matrix

There are wide variations among the thirty-one states of Mexico with respect to the percentage of the population living in urban areas (places with 15,000 or more population). As will be developed subsequently in this study, there is reason to believe that urbanization stimulates agricultural productivity and income. Consequently we believed it useful to analyze agricultural income in Mexico in more rural and more urban states. To do this we divided the thirty-one states into a group of fifteen more rural states and sixteen more urban states. The percentage of the population urban in the more rural states ranged from .1 to 21.5. The average was 14.8. The percentages of the population urban in the more urban states ranged from 25.6 to 74.3. The average was 42.3.

Table 9. Means and Coefficients of Variation of Variables Associated with State Agricultural Income Per Worker, 1960.

Variable	Rural states		Urban states	
	Mean	Coefficient of variation	Mean	Coefficient of variation
AGINC60	3,741	.14	5,313	.28
RURLF60	217,990	.64	136,720	.80
TRCPH60	1.09[a]	.71	3.98[a]	.76
IRRPL60	.053	1.14	.810	1.40
EDTPP59	6.89	.24	9.25	.21
PPLIT60	52.5	.18	71.5	.11
PABEJ60	47.0	.30	43.0	.36
Average % State Population Urban 1960	14.8		42.3	

[a]Number of tractors per 1000 hectares of arable land.

Table 9 shows means and coefficients of variation for the several variables associated with agricultural income in the two groups of states. As expected, the more urban states have higher agricultural income per worker than the more rural states. They also have more tractors per hectare, more irrigated land per worker, more primary teachers per capita, higher literacy rates, a smaller percentage of arable land in ejidos, and less agricultural labor.

Table 10. Correlation Matrix of Variables Assumed Linearly Associated with State Agricultural Income, Per Worker, 1960.

	AGINC60	RURLF60	TRCPH60	IRRPL60	EDTPP60	PPLIT60	PABEJ60
AGINC60	1.00						
RURLF60	-.39	1.00					
TRCPH60	.77	-.26	1.00				
IRRPL60	.78	-.18	.76	1.00			
EDTPP60	.45	-.67	.36	.22	1.00		
PPLIT60	.68	-.57	.60	.45	.72	1.00	
PABEJ60	-.27	.01	-.22	-.18	.34	-.21	1.00

The relationships among the variables shown in table 9 are reflected in simple correlations with state agricultural income, shown in table 10. The table indicates collinearity among some of the independent variables, suggesting that the simple correlation coefficients between these variables and agricultural income are biased. Strong collinearity could invalidate the use of multiple regression analysis of the relationships between the independent variables and agricultural income. In the present case, however, the collinearity does not appear that strong.[31]

Hypothesis Tests Via Multiple Regression

Table 11. Least Squares Regressions of 1960 Agricultural Income Per Worker Against 1960 Values of Independent Variables, Log and Linear, All States

	Log		Linear	
	Regression coefficient	t-value	Regression coefficient	t-value
CONSTANT	.815	8.27***	2960.	2.13**
RURLF60	-.080	-1.83**	-1.08	-0.86
TRCPH60	.005	1.48*	100.	1.10
IRRPL60	.032	2.14**	710.	3.08***
EDTPP60	-.040	-0.20	-36.	0.33
PPLIT60	.363	1.55*	31.2	1.69*
PABEJ60	-.185	-2.30**	9.2	0.87
R^2 / F	.70	9.34***	.77	13.05***
No. obs.	31		31	

Number of asterisks indicates significance level;

*10% **5% ***1% level.

[31]It is noted in N. H. Nie, et al, Statistical Package for the Social Sciences (McGraw Hill, New York, 1975) p. 340, that intercorrelations in the .8 to 1.0 range are indicative of extreme collinearity. However, L. R. Klein, An Introduction to Econometrics (Prentice-Hall, Englewood Cliffs, 1972) says that "Multi-collinearity is not necessarily a problem unless it is high relative to the overall degree of multiple correlation among all variables simultaneously. Production functions with overall correlations much in excess of .95...can be well estimated with inter-correlations...as high as .8 or .9".

The regression using the logs of the variables explains 70 percent of the variance in 1960 state agricultural income per worker, and is highly significant ($F = 9.34$, significant at the .01 level). Noting that the regression coefficient for RURLF $= -.08$, the elasticity of production of labor is, from equation (4'), $b_1 = 1 - b_3 - .08$. Since the regression coefficient for IRRPL (= .032) is the estimate for b_3, $b_1 = 1.0 - .032 - .08 = .888$. The elasticity of productivity of tractors is .005 and of irrigated land is .032

A one-tail test is used to test significance of the regression coefficients since the signs of the coefficients were included among the hypotheses, with the exception of PABEJ60 for which a two-tailed test is used. The coefficients for labor and irrigation are significant at the 5% level; that for tractors is significant at the 10% level. The effect of teachers per capita on agricultural income per worker is negative but insignificant; the literacy rate, however, is positively associated with agricultural income, as expected, and is significant at the 10% level. The structure of intercorrelations shown in table 2 explains these seemingly contradictory results. The correlation between primary school teachers per capita and agricultural productivity is positive and significant ($r = .45$). The correlation between literacy and productivity is higher ($r = .68$), perhaps reflecting the fact that literacy (as a proxy for knowledge) would directly affect agricultural productivity while number of teachers (a proxy for <u>future</u> knowledge) would have an indirect, or delayed, effect. The plausibility of this argument enhanced by the high correlation between literacy and the number of teachers ($r = .72$). When the two correlates are included as independent variables in the same regression, the conditional relation between teachers per capita (EDTPP) and the dependent variable is eliminated due to collinearity. Nevertheless, the results are consistent with primary school teachers being indirectly associated with agricultural productivity, literacy being the intervening variable.

The statistical findings concerning the contribution of the *ejido* form of agricultural organization are negative, and significantly so (t = 2.3). It might be argued that this reflects the less urban character of states in which *ejidos* are more important (see table 9). However, the relationship between urbanization and *ejidos* does not appear strong, and in any case, a regression that was run which controlled for urbanization still produced a significantly negative coefficient for *ejidos*.

Although not shown in the material presented so far, states in which *ejidos* are most important also have fewer tractors per hectare and smaller amounts of irrigated land per worker. However, this cannot explain the negative relationship between *ejidos* and agricultural income per worker because these capital variables are included in the regression shown in table 11 (TRCPH and IRRPL).

The results of the linear form of the regression show the same direction of relationship as those found in the logarithmic form. Although the linear form provides better explanatory power than the log form (R^2 = .77 versus .70), only two variables in the linear form are significant: irrigation (at the 1 percent level), and literacy rates (at the 10 percent level). The high predictive power yielded by the linear relation between income per worker and irrigated land per worker (t = 3.08) suggests a more complicated agricultural income equation than either the log or linear forms tested here. We return to this possibility in the last section of this chapter.

Effects on State Agricultural Income of Out-Migration and Urban Wages

If there is interdependence between agricultural income per worker, rural-to-urban migration, and urban wages, then the use of ordinary least-squares regression to estimate equation (4') is not strictly legitimate. Such interdependence suggests that equation (4') should be modified by the inclusion of an out-migration variable and urban wages, and estimated with a simultaneous equations estimator.

Changes in agricultural income per worker change the attractiveness of rural areas relative to urban areas, thus affecting rural-to-urban migration. However, migration causes changes in the supply of agricultural labor, thus, "feeding-back" on agricultural income. Given this feed-back effect, neo-classical theory predicts that states with higher rates of out-migration will have higher agricultural income per worker, other things the same.

However, migration may also affect the quality of the agricultural labor force, if more able workers have a higher propensity to migrate than less able workers. In this case, states with higher rates of migration could have lower agricultural income per worker if the effect of migration in reducing labor quality is greater than its effect in reducing labor supply.

We argue that the level of urban wage in a state is a proxy for two factors affecting agricultural income per worker in that state: (1) urban demand for agricultural commodities, and hence for agricultural labor; (2) the transmission of agricultural knowledge and inputs not included in our agricultural production function. For example, with respect to the first factor, higher urban wages reflect higher per capita income in urban places, indicating that high-wage urban areas ought to have higher demand for agricultural commodities than low-wage urban areas. On this score, therefore, we would expect states with high urban wages to also have high agricultural productivity and income per farm worker.

The effect of migration on agricultural income can be estimated by including the state-to-city migration rate in the agricultural income equation. The migration rate (OMRLF) will be specified as 1960-1970 state-to-city migration relative to the 1960 state agricultural labor force.[32] However, since agricultural income may be a determinant of state-to-city migration (and is so specified in chapter 4), two-stage least regression will be used for this estimation to reflect the simultaneous determination of agricultural income and migration. Because the migration rate may have both positive and negative effects on agricultural income per worker, we make no hypothesis about the sign of OMRLF in the regression equation.

We also include the state urban wage (AUWAG) in the regression to capture the two effects, noted above, that the urban wage may have on agricultural income per worker. Since agricultural income per worker is specified in chapter 5 as a determinant of the urban wage, two-stage least squares regression is appropriate to reflect the simultaneity between these two variables. For each state AUWAG is measured as the weighted average wage in 1965 of the cities in that state with

---

[32]For 1960-1970 the migration data show movement from each of the 31 states to each of 36 cities with more than 50,000 population in 1960. More detail on the migration data is given in chapter 4 and in the appendix on data sources.

populations of more than 50,000 in 1960. Weights are the labor force of each included city.

Since both effects of the urban wage on agricultural income per worker are hypothesized to be positive, we expect AUWAG to have a positive sign in the regression.

Table 12 shows simple correlations between the two new variables (OMRLF and AUWAG) and the other variables included in the two-stage least squares regression. Table 12 can be thought of as a correlation submatrix to be appended to table 10. Teachers per capita (EDTPP) is excluded from table 12 because of its high correlation with literacy (PPLIT) noted earlier.

Table 12. Simple Linear Correlations of OMRLF and AUWAG with Variables Specified in the AGINC Regression

Variable	OMRLF	AUWAG
AGINC60	.07	.32
RURLF60	.33	.37
TRCPH60	.25	.03
IRRPL60	.16	.20
PPLIT60	.03	.19
PABEJ60	.07	-.37
OMRLF60	1.00	.18

The results of the two stage least squares regression including OMRLF and AUWAG are shown in table 13. Given the theoretical and statistical attributes of the rural productivity specification, only the log-linear equation is estimated.

After adjustment for the inflated degrees of freedom (see note to table 13), all variables are significant at the 5 percent level or less with the exception of PABEJ, and the R^2 is highly significant. The coefficient for the average urban wage (AUWAG) has the expected sign. We interpret the negative sign of the migration rate (OMRLF) to mean that the negative effect of migration in lowering the quality of the agricultural labor force more than offsets its positive effect in reducing the supply of agricultural labor.

Table 13. Two-Stage Least Squares Log-Linear Regressions of Agricultural Income Per Worker on Independent Variables

	Regression Coefficient	t-value
CONSTANT	.449	17.23 ***
RURLF	-.067	-10.70 **
TRCPH	.010	20.51 ***
IRRPL	.014	9.03 **
PPLIT	.359	18.00 ***
PABEJ	-.023	- 1.06
OMRLF	-.001	- 2.84 ***
AUWAG	.320	12.56
R^2 / F	.84	371.2
n	485	

Note: Only 26 states were included in this regression; five states were deleted that did not contain cities with more than 50,000 inhabitants in 1960. For states with more than one such city, a weighted average of AUWAG was used. Hence the number of observations on AUWAG is less than 36, the number of cities with more than 50,000 inhabitants in 1960. The n of 485 reflects the number of times each state was repeated due to state-to-city migration flow. This greatly overstates the true degrees of freedom for all variables except OMRLF, the migration variable. To correct for this the standard error of the regression coefficient of each of the affected variables was multiplied by $[(n-k-1)/(n-k-1)]^{\frac{1}{2}}$ where n is the number of states (the true number of degrees of freedom) and k is the number of variables. The corrected significance level is indicated by the number of asterisks, following the format of table 11.

The two-stage least squares regression explains more of the variance in state agricultural income per worker than the ordinary least squares regression shown in table 11 (for which $R^2 = .70$ in the log-linear form). This suggests that there is significant interdependence between agricultural income on one hand and migration and urban wages on the other, and that the two-stage least squares regression, consequently, is a better predictor of state agricultural income per worker than the ordinary least squares regression.

The inclusion of OMRLF and AUWAG and use of the two-stage regression also affects the coefficients on the other variables included in the analysis, except literacy (compare tables 13 and 11). The significance of the differences in the coefficients is not easy to determine. Part of the differences may reflect the fact that the data sets underlying the two regressions are not the same. The table 11 regression is based on data for 31 states. For the reason given in the note to table 13, the regression in that table is based on data for 26 states. The relatively small difference between the coefficients for size of agricultural labor force (RURLF) perhaps can be explained by this difference in data sets.

A different, or additional, explanation probably is needed, however, to account for the differences between the coefficients for tractors per hectare (TRCPH), irrigated land per worker (IRRPL) and percent of arable land in _ejidos_ (PABEJ). In the case of tractors, some of the difference may be explained by the positive correlation between tractors and the migration rate (see table 12). States with higher migration rates tend to have more tractors per hectare, perhaps reflecting substitution of tractors for labor. Consequently the coefficient for tractors in table 11 may be picking up some of the negative effect of the migration rate on agricultural income per worker (see table 13). When the migration rate is entered separately in the regression, as in table 13, its negative effect on the tractor coefficient is removed and the latter rises, as the comparison of tables 11 and 13 indicates.

It seems more difficult to make the same argument to explain the difference between the coefficients for irrigated land per worker in tables 11 and 13. The correlation of this variable with the migration rate is weaker than in the case of tractors and is positive (see table 12). Since the irrigation variable is sharply lower in table 13 than in table 11, the correlation with the migration rate would explain this behavior only if the correlation were negative and probably much stronger than is shown in table 12.

The correlation between the irrigation variable and the urban wage is only slightly stronger than it is between irrigation and the migration rate, but it is positive in the former case. Consequently, the irrigation coefficient in table 11 may be picking up some of the positive influence of the urban wage on agricultural income. In this case, when the urban wage is entered separately, as in table 13, the coefficient for irrigation would decline, as it does. However, the decline seems sharp (55 percent) relative to the modest correlation between the urban wage and irrigation (r = .20, table 12). We conclude that we have no satisfactory explanation for the difference between the coefficients for irrigation in tables 11 and 13.

The most marked difference between the two tables is in the coefficient for *ejidos*. The coefficient in table 13 is only 12 percent of that in table 11. The sign of the coefficient is the same in both cases (negative), but in table 13 the coefficient is not significant.

The negative correlation between the *ejido* variable and the urban wage (see table 12) perhaps explains some of the difference between the *ejido* variable in tables 11 and 13. The correlation indicates that states with larger percentages of arable land in *ejidos* had lower urban wages. Since the urban wage had a strong positive relation to agricultural income (table 13), some of the negative relation between *ejidos* and urban wage would be picked up in the negative relation between *ejidos* and agricultural income. When the urban wage is entered separately in the agricultural income regression, as in table 13, the effect of the negative *ejido*--urban wage relation on the *ejido*--agricultural income relation is eliminated. In this case, we would expect the *ejidos* coefficient in table 13 to be less than in table 11.

Hypothesis Testing of Irrigation Impacts with Simple Regression

In the discussion of table 10 we noted that there appeared to be strong collinearity among some of the independent variables. This is particularly true of irrigated land per farm worker (IRPL) and tractors per hectare (TRCPH). Between these two variables $r = .76$. While we concluded that the collinearity between these variables, and among them and others, did not invalidate a multiple regression analysis of agricultural income, the collinearity may nevertheless be sufficient to bias the coefficients shown in tables 11 and 13.

We decided it would be useful to experiment with a simple regression of agricultural income per man on irrigated land per man. The rationale is that irrigation not only is correlated with tractors per hectare, it likely is correlated also with other important determinants of agricultural income per worker for which we lack data, such as fertilizers, pesticides, and improved seed varities. There is much evidence that in the development of Mexican agriculture there has been strong complementarity between irrigation and these other inputs, as well as with tractors.[33] To the extent that this is true, irrigation can be taken as a proxy for an entire technology, the components of which are irrigation water and the complementary inputs.

Based on this argument we estimated the simple regression equation shown in table 14. The results support the hypothesis that irrigation represents a cluster of inputs that together constitute what we will call an irrigation technology. In states which had some amount of irrigation, this technology accounted for two-thirds of the variation among states in agricultural income per man (r^2 = .68). Obviously other factors also were involved. Many of the states had small amounts of irrigated land. In those states the irrigation technology would not likely be the dominant factor in agricultural production. Even among states where most agriculture was irrigated, we still would expect variations in output per man

---

[33]Reed Hertford, Sources of Change in Mexican Agricultural Production, 1940-1965 (U.S. Department of Agriculture, Foreign Agriculture Report Number 73, 1971). Also, with respect to agriculture in Northwest Mexico, see Pierre Crosson, "Urban-Rural Relationships in the Northwest Region," chapter 3 in Ronald Cummings, Interbasin Water Transfers: A Case Study of Mexico (Baltimore, Johns Hopkins University Press for Resources for the Future, 1974).

since it is unlikely that irrigated land and other components of irrigation technology are perfect complements. Nonetheless, the evidence from table 14 and from tables 11 and 13 strongly supports the argument that irrigation technology, represented in our analysis by the amount of irrigated land per farm worker, was a powerful determinant of agricultural income per worker.

Table 14. Simple Linear Regression of Agricultural Income Per Man on Irrigated Land Per Man in 1960

	Regression coefficient	
	All states	Excluding states lacking irrigation[a]
Constant	4022	3917
IRRPL	1177	1223
r^2	.61*	.68*

[a]Five of Mexico's states had no irrigated land in 1960. Income is in pesos of 1960.

*Significant at 1 percent level.

Appendix to Chapter 3

The logarithmic derivation of equation (4') with respect to time gives an equation that is linear in the growth rates of all variables. For policy analysis, this way of expressing the relationship between agricultural income per man and other variables probably is preferable to the forms underlying tables 11 and 12. This is because an equation based on growth rates in the variables would reflect the response of agricultural income _through time_ to changes in the independent variables. The implementation of policies and achievement of results of course take time. By contrast, tables 11 and 12 depict relationships at a _point in time_. From these tables one can say what agricultural income would have been at that point in time _if_ one or more of the independent variables had had a different value.

We experimented with a form of equation (4') which was linear in the growth rates of the variables. The results are shown in table 15. We had 1950 data for only 21 states and no 1950 data for irrigated land per worker for any state. Consequently, there are only 15 degrees of freedom in the equation in table 15 compared with 24 in the equation in table 11 and 29 in the equation in table 14.

The equation in table 15 gives poor results. We believe the fewer degrees of freedom plus the absence of data on irrigated land per worker, a strong variable in table 11 (the only independent variable in table 14) may account for this. In any event, the "growth rate" form of the agricultural income equation does not tell us enough about the determinants of agricultural income to be useful for policy analysis. Consequently, in our discussion of policy in chapter VI, we rely on the "point-in-time" form of the agricultural income equation.

Table 15. OLS Regression of 1950-1960 Growth Rate of AGINC Against 1950-1960 Growth Rates of Independent Variables

	With EDTPP included		Without EDTPP	
1950-1960 % increase	Regression coefficient	t-value	Regression coefficient	t-value
CONSTANT	.937	1.41	.718	1.29
RURLF	-3.137	-1.62*	-2.680	-1.51*
TRCPH	.043	0.42	0.058	0.59
PPLIT	10.390	2.57**	9.558	2.54**
PABEJ	-0.109	-0.11	-0.336	0.37
EDTPP	-0.883	-0.065	--	--
R^2 / F	.34	1.58	.33	1.94
No. obs.	21		21	

Chapter 4

DETERMINANTS OF RURAL TO URBAN MIGRATION

Specification of the Migration Equation

We hypothesize that the gross migrant flow M between two locations i and j depends upon the individual's perception (K) of the expected economic rate of return (P) that could be gained by moving from i to j, the individual's propensity to respond (Π) to these perceptions, and the "at risk" population at i available to migrate (L).[34]

$$M_{ij} = f(P_{ij}, K_{ij}\Pi_i, L_i). \quad (6)$$

The variables that determine the expected rate of return (P_{ij}) are the annual wages (W_j) in the place of destination; available public goods in that place (G_j); the individual's expected remaining years of working life as indicated by age (A_i); the probability of employment at the place of destination, as indicated by the unemployment rate (U_j) earnings foregone at

---

[34]The relevant literature is extensive. We have found the following particularly useful: Daniel J. Hogan and Manuel T. Berlinck, "Conditions of Migration, Access to Information, and First Jobs: A Study of Migrant Adaptation in Sao Paulo, Brazil," in A. H. Richmond and D. Kubat, eds., Internal Migration: The New World and the Third World (Beverly Hills, Calif., Sage Publications, 1976); A. Schwartz, "Interpreting the Effect of Distance Migration," Journal of Political Economy, vol. 81 (1973); Arthur L. Silvers, "Probabilistic Income Maximizing Behavior in Regional Migration," International Regional Science Review, vol. 2 (Fall 1977); R. A. Hart, "Interregional Economic Migration: Some Theoretical Considerations" (Part II), Journal of Regional Science, vol. 15 (1975); B. Herrick, "Urbanization and Migration in Latin America: An Economist's View," in F. Rabinowitz and F. Trueblood, eds., Latin American Urban Research, vol. 1 (Sage Publications, 1971); L. Sjaastad, "The Costs and Returns, of Human Migration," Journal of Political Economy, vol. 70 (1962); M. Todaro, "A Model of Labor Migration and Urban Unemployment," American Economic Review, vol. 69 (1969); M. J. Greenwood, "Research on Internal Migration in the United States: A Survey," Journal of Economic Literature, vol. 13 (1975); K. G. Willis, "Problems in Migration Analysis," D. C. Health (1974).

the place of origin i, which are a function of the values of these same variables at i; and moving costs that increase with distance (D_{ij}).

$$P_{ij} = r\ (W_i, W_j, G_i, G_j, U_i, U_j, A_i, D_{ij}) \qquad (7)$$

Although it may be true that the returns from migration to a given location are substantial, the potential migrant may lack the information needed to perceive this. Accurate perceptions (K) of the returns depend upon the accuracy and completeness of labor market information received at i from locations j. We hypothesize that this information improves with the number and quality of transmitters at j and receptors at i, and declines as transmission distance increases. The variables used as proxies to measure these concepts are the number of labor market participants (L) at both i and j, the number of participants once located in i but who, due to past migration ($M_{t-\tau}$), are now located in j, and the distance from i to j.

$$K_{ij} = k(L_i, L_j, M_{ijt-1}, \ldots, M_{ijt-n}, D_{ij}) \qquad (8)$$

Assuming that the individual at i perceives a gain from migration, migration may still not occur if the individual resists relocation or merely delays a response. We hypothesize that the migration response propensity (Π) is a function of income, age, and other socio-demographic characteristics (S) of the "at risk" population at i

$$\Pi_i = p(W_i, A_i, S_i). \qquad (9)$$

Substituting the relations specified in (7), (8), and (9) into equation (6), the gross migrant flow from i to j is

$$M_{ij} = f[r(W_i, W_j, G_i, G_j, U_i, U_j, D_{ij}, A_i); k(L_i, L_j, D_{ij}, M_{ij-1}, \ldots, M_{ijt-n}); p(W_i, A_i, S_i) ; L_i] \quad (6')$$

If the migration function take the form of the gravity model,[35] then the relation will be multiplicative, written as

$$M_{ijt} = aW_i^{r_1+p_1} W_j^{r_2} G_i^{r_3} G_j^{r_4} U_i^{r_5} U_j^{r_6} L_i^{1+k_1} L_j^{k_2} D_{ij}^{r_7+k_3} A_i^{r_8+p_2} S_i^{p_3} \prod_{\tau=1}^{n} M_{ijt-\tau}^{k_{4+\tau-1}} \quad (10)$$

where the parameters r_1, r_3, r_6, r_8, and p_2 are negative, p_1, r_2, r_4, and r_5 are positive; and k_1, k_2, k_3, and $k_{4+\tau}$, (τ=1,..., n)are either positive or negative, depending upon whether the information effect of the associated variable reflects uncertainty or search costs.[36]

Definition and Measurement of Variables

The analysis is focused on explanation of state-to-city migration in Mexico from 1950 to 1960 and from 1960 to 1970. The data were taken from

---

[35]The gravity model as used in regional analysis makes use of an analogy with Newtonian physics: the strength of the attraction between two population centers is assumed to vary directly with the product of their size and inversely with the square of the distance between them. See W. Isard, Methods of Regional Analysis (Cambridge, Mass., MIT Press, 1960).

[36]See A. L. Silvers, "Probabilistic Income Maximizing Behavior in Regional Migration," International Regional Science Review (April 1977).

the population censuses of 1950, 1960, and 1970, as well as from other sources.[37] With some exceptions, noted below, other independent variables for predicting migration were measured as of 1950 (for 1950-1960 migration) and 1960 (for 1960-1970 migration). For 1950 to 1960 the data relate to migration from each of 21 states to each of 36 cities with populations of 50,000 or more in 1960. For 1960 to 1970 the data relate to migration from 31 states to each of the same 36 cities. For each period the data count people who arrived in the city of destination in the previous decade.

There are a number of characteristics of the migration data which should be understood in considering the results presented in this chapter.

1. The data do not necessarily measure the movement of people <u>directly</u> from the state where they were born to the city where they were living when the census counted them. In the interim they may have first moved to one or more other states, then have come to the city where they were counted.

2. The data count <u>all</u> of the people born in a given state who moved to a given city in the preceding decade, not just those from rural areas.

3. The data do not count movement within states, i.e., people who were born in the state where the city is located and who moved to the city during the decade are not counted.

Our analysis postulates that certain economic, social, and demographic variables in the state of birth in the decade when migration to the city

---

[37]See the appendix at the end of the paper for data sources.

occurred help to explain the number of people born in that state who made the migration. But if the preponderence of migrants moved from states other than the one in which they were born, then taking account of state-of-birth variables according to our postulate will fail to explain the observed patterns of moves. Our data, it is emphasized, measure migrants according to state-of-birth rather than the state from which they moved.

In the belief that for the more rural states in Mexico, the state of birth was the state from which the great majority of migrants moved, we sought to mitigate the measurement problem by separating the rural states from the more urbanized states and running separate migration analyses. In this case, the rural state analysis would be expected to yield a more satisfying degree of explanation.

However, it may be that many potential migrants living in states other than the one in which they were born remain sensitive to economic events in their state of birth. This, no doubt, describes the behavior of return migrants. In this case, data such as ours showing migration by state of birth is appropriate.

There are two reasons for believing that the potential measurement problem with our migration data is not serious. First, while our results are not uniform in this respect, in general they show a strong statistical relationship between conditions in state i and measured migration to city j. In some of the regressions all of the state-of-birth variables are significantly related to measured migration, and in all of the regressions some of these variables are significant. The best fits to the data were obtained when the data were expressed in logarithms, and all of the logarithmic regressions had R^2s of .85 to .87. This was true of regressions for more urban states which, being more urban, might be expected to have a relatively large number of people who migrated there from some other state, as it was of the more rural states.

These results suggest that although the migration data do not necessarily show direct movement within the decade from state-of-birth to city-of-destination, most of the moves in fact were of this sort. The alternative explanation is that the relevant state-of-birth variables were highly correlated with the relevant, and possibly not the same, state-of-origin variables. Of course, we cannot disprove the alternative explanation, but we consider it less plausible than the first.

The strength of the regressions relating state-of-birth conditions to measured migration is the strongest argument for the validity of the migration data. There is, however, a second reason for believing that the problems with these data are not serious, namely evidence that return migration is an important part of the migratory process.[38] If return migration is important, then conditions in state i would be relevant not only to people living in i at time t but also to all those born in i but who at time t lived in other states.

Indeed, the possibility that return migration is important indicates that for our purposes migration data by state-of-birth may be preferable to data by state-of-immediate-origin. One of our principal interests is to examine the effects of increased rural investment and income in state i on urban-bound migration. If return migration is important, the increase in income in state i not only would reduce direct out-migration from the

[38] J. Vanderkamp, "Return Migration: Its Significance and Behavior," Western Economic Journal, vol. 10, 1972, pp. 460-465. J. B. Kau and C. Sirmans, "New, Repeat and Return Migration: a Study of Migrant Types," Southern Economic Journal, vol. 43, Oct. 1976, pp. 1144-1148. Kau and Sirmans, "The Influence of Information Cost and Uncertainty on Migration: a Comparison of Migrant Types," Journal of Regional Science, vol. 17, April, 1977.

state to urban areas in other states, it also would induce some natives of state i now living elsewhere to return to the state rather than migrate to urban areas in other states. Migration data reflecting state-of-birth thus better represent the total population likely to respond to an increase in income in state i than data reflecting state-of-immediate-origin.

Since the migration data count all the people who moved, not just those from rural areas, we of course do not claim to be analyzing only rural-to-urban migration. However, we are particularly interested in rural-to-urban migration, and the specification of our regression models includes variables we think relevant to that sort of movement. Most of the discussion of this chapter is focused on the role of those variables, particularly average agricultural income, and we believe we have achieved significant insights concerning its role in explaining migration. The implication is either that most migration in fact was from rural areas or that conditions in urban areas in state of birth affecting migration were highly correlated with the relevant conditions in rural areas in those states. We find in fact that average agricultural incomes and urban wages in states of birth were correlated ($r = .71$, see AGINC and CWAGE in table 25 of the next chapter). This suggests that most of the migration from states could be from urban rather than rural areas, and that the observed relations between migration and agricultural income simply reflects the correlation with the state urban wage. While this is the possible explanation, we think it implausible. Most of the people in Mexico in the period studied lived in rural areas (see table 9). Moreover, the size of the state agricultural labor force was positively correlated and highly significant in explaining migration in all but one of our regressions (see tables 18-22), and size of agricultural labor force in the state was negatively correlated with the state's urban labor force. This supports the inference that in fact much, if not most, of the observed migration was from rural areas.

Our inability to measure intra-state migration to cities of course means that we have not accounted for all internal migration to urban areas in the 1950s and 1960s. The number of people who thus escape our net probably is significant since distance is strongly and negatively related to migration (see SCDIST in tables 18-22). However, unless the determinants of intra-state migration are significantly different from those of interstate migration--and we see no reason to believe that they are--the exclusion of intra-state migration should not bias our results. It is worth noting in this connection that in the 1950s and 1960s almost half of the migration to urban areas in Mexico was to the metropolitan are of Mexico City (see table 4). Since we treat this entire area as a city-of-destination, there is no intra-area migration there. That is, <u>all</u> migration to the Mexico City area is from the rest of the country (except for a trivial number of internatonal migrants), hence is included in our analysis.

With these caveats spelled out, let us proceed.

<u>Dependent variable</u>

$SCMIG_{jt}$: The number of people who migrated to city j who were born in state i, 1950 to 1960 and 1960 to 1970. Represents M_{ijt} in equation (10).

<u>Independent variables: economic rate of return</u>[39]

$CWAGE_{jt}$: Average annual wage in service industries, in pesos, in city j in 1955 (assumed relvant to migration from

---

[39]There are no data corresponding to variables G_j and U_i in equation (10).

1950 to 1960) and 1965 (assumed relevant to migration from 1960 to 1970). Represents W_j in equation (10).

$AGINC_{it}$: Annual labor income per worker in agriculture in state i in 1960, assumed relevant to migration from 1960 to 1970. Comparable data for 1950 were not available, so we substituted agricultural output per person in the agricultural labor force. These variables represent W_i in equation (10).[40]

$GHEPP_{it-1}$: Per capita federal government expenditures on health in state i in 1959-1963, assumed related to migration 1960-1970. Comparable data relevant to migration 1950-1960 not available. Represents G_i in equation (10).

$GEDPP_{it-1}$: Per capita federal government expenditures on education in state i in 1959-1963, assumed related to migration in 1960-1970. Comparable data relevant to 1950-1960 migration not available. Represents G_i in equation (10).

$EDTPP_{it-1}$: Primary school personnel per capita in state i in 1949 (assumed related to 1950-1960 migration) and

---

[40] Since the migration data are total state-to-city migration, not migration only from rural areas, an origin income variable reflecting earnings of both rural and urban workers would have been preferable. However, most of the people in Mexico lived in rural areas in both 1950 and 1960 (see table 1), suggesting that agricultural income is an appropriate indicator of state wages. Moreover, table 13 (chapter III) indicates that the urban wage in the state of origin (AUWAG) was highly correlated with agricultural income in that state.

1959 (assumed related to 1960-1970 migration). Represents G_i in equation (10).

$CLMPR_{jt-1}$: Natural population increase in city J from 1940 to 1950, divided by city j labor force in 1950, assumed related to migration 1950-1960. A comparable figure was used for migration 1960-1970. The higher this ratio, the greater the contribution of natural population increase in one decade to growth of the urban labor force in the succeeding decade. Hence this variable reflects the pressure of natural population increase on the employment absorption capacity of the urban labor market. It is assumed to be positively correlated with urban unemployment and represents U_j in equation (10).

Independent variables: information and response propensity

$CLAFO_{jt-1}$: The labor force in city j in 1950 (assumed related to migration 1950-1960) and 1960 (assumed related to migration 1960-1970). Represents L_j in equation (10).

$SLAFO_{it-1}$: The agricultural labor force in state i in 1950 (assumed related to migration 1950-1960) and 1960 (assumed related to migration 1960-1970). Represents L_i in equation (10).

$SCDIS_{ij}$: Highway distance in miles from the principal city of state i to city j. Represents D in equation (10).

$PPYNG_{it-1}$: Percent of the population age 10-29 in state i in 1950 (assumed related to 1950-1960 migration) and 1960 (assumed related to 1960-1970 migration). Represents A_i in equation (10).

$PPLIT_{it-1}$: Percent of the population, age six or more, literate in state i in 1950 (assumed related to 1950-1960 migration) and 1960 (assumed related to 1960-1970 migration). Represents S_i in equation (10).

$MIGST_{ijt-1}$: The number of people living in city j in 1960 who were born in state i. Assumed related to 1960-1970 migration. Comparable data for 1950 not available. Represents M_{ijt-1} in equation (10).

Estimation Problems

There are two problems affecting the estimation of the parameters contained in equation (10). The first is that migration to cities and urban wages may be simultaneously determined.[41] If so, the coefficient indicating the elasticity of migration with respect to the urban wage (r_2 in equation 10) would be subject to identification bias, if ordinary least squares regression were used for parameter estimation. Two-stage least squares regression is an appropriate procedure in this case.

---

[41] M. Greenwood, "Urban Economic Growth and Migration: Their Inter-Actions," Environment and Planning, vol. 5 (1973); R. Muth, "Migration: Chicken or Egg?" Southern Economic Journal, vol. 13 (1971).

A second problem is in the estimation of the information effect of past migration, i.e., the elasticities of migration with respect to the lagged migration stream ($k_4,\ldots,k_n$). Time series data for past migration are required for this estimation, but are not available. Instead we have used the migrant stock in each city classified by state of origin to represent this variable. In so doing, we assume that the elasticity parameters are the same for all past migrants regardless of the length of time since the migration was made, i.e., $k_4=k_5=\ldots=k_n$. We also assume that the information effect of i to j migrant populations for each prior time period is additive rather than multiplicative, meaning that prior i to j migrants can be added. The resulting sum, the total i to j migrant stock, can then be regressed against current i to j migration.[42]

Presentation and Discussion of the Data

Table 16 gives means and coefficients of variation of the variables associated with state-to-city migration. Data relevant to migration from 1950 to 1960 and 1960 to 1970 are shown, except that data relevant to the earlier period were not available for migrant stock (MIGST) or for per capita federal government expenditures in the states on education and health (GEDPP and GHEPP). The distance variable (SCDIST) is not dated since it is assumed to have been the same in both periods.

---

[42]An alternative is to use a Koyck-type distributed lag model in which the deteriorating effect of time on the flow of information is assumed constant. We experimented with such a model but found the results unsatisfactory. The details are in the appendix to this chapter. For an account of distributed lag models see L. M. Koyck, Distributional Lags and Investment Analysis (Amsterdam, North Holland, 1954); also, most modern econometrics texts include the treatment of distributed lags.

Table 16. Means and Coefficients of Variation of Variables Associated with State-to-City Migration

Variable	Units	Rural states Mean	Rural states Coefficients of variation	Urban states Mean	Urban states Coefficients of variation
SCMIG60	people	1,457	3.84	1,098	2.74
SCMIG50	people	1,511	4.79	1,465	4.10
SLAFO60	people	363,500	0.67	252,500	0.74
SLAFO50	people	212,100	0.57	171,900	0.81
CLAFO60	people	96,100	2.99	96,700	3.00
CLAFO50	people	58,300	2.93	58,600	2.94
CWAGE60	pesos	8,182	0.45	8,115	0.45
CWAGE50	pesos	5,276	0.42	5,279	0.42
AGINC60	pesos	3,741	0.13	5,313	0.28
AGINC50	pesos	1,764	0.45	3,126	0.30
GEDPP60	pesos	17.17	0.95	19.37	0.58
GEDPP50	pesos	38.12	0.64	107.06	0.70
SCDIST	miles	730.8	0.72	828.3	0.61
CLMPR60	ratio	0.807	0.16	0.807	0.16
CLMPR50	ratio	0.543	0.26	0.540	0.26
PPYNG60	percent	37.9	0.05	37.9	0.03
PPYNG50	percent	38.6	0.03	38.2	0.04
PPLIT60	percent	52.5	0.18	71.5	0.11
PPLIT50	percent	46.1	0.22	64.8	0.14
MIGST60	people	3,625	5.76	2,465	5.02

Note: Variables labeled 50 are relevant to migration between 1950 and 1960. Those labeled 60 are relevant to migration between 1960 and 1970.

As indicated in the discussion of table 9, the 31 states of Mexico were classified as rural or urban, depending upon the percentage of the population which was urban in 1960. However, for the analysis of migration between 1950 and 1960, we had data for only 21 states. This is because the data for 1950-1960 and 1960-1970 migration were collected in separate periods and our resources for data collection were more abundant in the second period than in the first.

Twelve of the 15 states considered rural in 1960 and 9 of the 16 considered urban were included among the 21 for which 1950-1960 data were collected.

Because the number of states is not the same in the two periods, the state data in table 16 are not strictly comparable between periods. For example, we cannot say that average annual agricultural income in the rural states increased from 1,764 pesos in 1950 to 3,741 pesos in 1960. Comparisons between rural and urban states in either 1950 or 1960 are not affected, however, only those between years.

In the subsequent analysis we sometimes use regressions based on pooled data for the two periods. Whenever that occurs we of course use data only for the 21 states represented in both periods.

The differences in table 16 between rural and urban states with respect to urban labor force (CLAFO), urban labor market pressure (CLMPR), and urban wages (CWAGE) reflect statistical anomalies, small enough to be ignored. Conceptually the urban data for a given year are the same for both sets of states since these data are for the same set of 36 cities.

Some additional comment on the data in table 16 will help in interpreting them. For example, between 1960 and 1970 an average of 1,457 people born in each of the 15 rural states moved to each of the 36 cities.

Total movement of such people to the cities in the decade, therefore, was 786,780. The comparable figure for movement of people born in the 16 urban states was 632,448. The labor force in the 36 cities increased 65 percent from 1950 to 1960 (from 58,300 to 96,100) and the urban wage increased 55 percent in nominal terms (from 5,276 pesos in 1950 to 8,182 pesos in 1960). The urban wage exceeded average agricultural income per worker in the rural states by a factor of 3 in 1950 and a factor of 2.2 in 1960. Average agricultural income in the urban states was higher than in the rural states in both years, but the difference narrowed substantially over the decade.

The federal government spent more on health (GHEPP) and education (GEDPP), per capita, in the more urban states than in the more rural states, the difference being particularly marked for health. We have no information on state government expenditures for these services, but public finance in Mexico is strongly centralized in the federal government.

The ratio of natural population increase in the 36 cities to city labor force was much higher in 1960 than in 1950. As indicated in the definitions of variables, this ratio is a proxy for pressure on the urban labor market owing to natural population increase. Our hypothesis is that city natives are more familiar with the local urban market than potential migrants, thus giving the natives an advantage in that market relative to migrants. If this is correct, then the higher this ratio, the less attractive the urban labor market to the potential migrant.

Finally, migrants from the more rural states in the 1960s had substantially more predecessors to follow, as indicated by the migrant stock (MIGST), than did migrants from the more urban states. This indicates that the higher rates of migration from the more rural states in the 1950s and 1960s likely was characteristic of earlier decades also.

Table 17 gives the matrix of simple correlations for migration from 1960 to 1970. Few of the variables are strongly correlated with migration. The rate-of-return variables in states of origin, that is, agricultural income (AGINC), and government expenditures on health and education (GHEPP and GEDPP), all have the expected negative signs, as does the variable for urban labor market pressure (CLMPR). The urban wage (CWAGE), urban labor force (CLAFO), agricultural labor force in state of origin (SLAFO), and the migrant stock (MIGST) have positive signs, also as expected. Note that the migrant stock, reflecting cumulative migration, has the same correlation pattern with the other variables as does migration from 1960 to 1970. This suggests stability over time in the relationships between migration and factors affecting it and supports the assumptions underlying the use of migrant stock (MIGST) in place of lagged migration levels ($MIG_{ijt-\tau}$; $\tau = 1, \ldots, n$); see page 69.

The correlation matrix indicates a strong relationship ($r = .68$) between state literacy (PPLIT) and agricultural income (AGINC) and between government expenditures on health in the states (GHEPP) and agricultural income ($r = .78$), suggesting that there may be problems of collinearity between these variables. This issue is considered later in discussion of the regression analyses of the data.

Because the urban wage and migration to cities may be simultaneously determined, all regressions in this chapter are estimated using two-stage least squares. The values of the urban wage (CWAGE) used to estimate the coefficients of CWAGE in the migration equation are <u>estimated</u> values derived together with the urban wage equation described in the next chapter.

Table 17. Correlation Matrix of Variables Assumed Linearly Associated with Migration from 1960 to 1970.

	SCMIG	CWAGE	AGINC	GEDPP	GHEPP	CLAFO	SLAFO	SCDIST	CLMPR	PPYNG	PPLIT	MIGST
SCMIG60	1.00											
CWAGE60	.27	1.00										
AGINC60	-.06	-.03	1.00									
GEDPP60	-.09	-.00	.18	1.00								
GHEPP60	-.07	-.02	.78	.06	1.00							
CLAFO60	.57	.29	.00	.00	.00	1.00						
SLAFO60	.16	.01	-.26	-.48	-.39	.00	1.00					
SCDIST	-.18	.39	.28	.30	.24	.09	-.19	1.00				
CLMPR60	-.17	-.46	.01	-.00	.00	-.21	.00	.17	1.00			
PPYNG60	-.01	.00	-.03	-.17	-.10	.00	.04	.08	-.00	1.00		
PPLIT60	-.04	-.01	.68	.27	.57	.00	-.40	.13	.00	-.08	1.00	
MIGST60	.83	.20	-.06	-.06	-.06	.59	.12	-.14	-.13	-.06	-.05	1.00

The Basic Migration Model

Table 18 presents regression coefficients and t-values for both linear and logarithmic regressions with migration from 1960 to 1970 (SCMIG60) as the dependent variable, and with the variables shown in the correlation matrix (table 16) as the independent variables.[43]

The significance tests are either one or two-tailed, depending upon whether the sign of the variable was unambiguously included in the specification of hypotheses in the first section of this chapter.

In the logarithmic regression shown in table 18, destination city wage CWAGE60 is positively and significantly associated with migration as expected, and destination city labor market pressure (CLMPR) is significant at the .05 level and negative as expected.

In equation (10) we specified that the coefficient of AGINC (W_i in equation (10)) has two components: r_1, representing the wage in i and expected to be negative; and p_1 representing ability to pay moving costs and "propensity to move," and expected to be positive. If this specification is correct, the positive and significant coefficient for AGINC in table 18 means that the p_1 component of the coefficient is greater than the r_1 component.

The t value for the coefficient of GEDPP is 2.92; however, GEDPP is a "rate of return" variable in the place of origin (corresponding to G_i in equation (10)) and so is expected to have a negative sign. (Its

[43]As indicated above CWAGE60 is also an endogenous variable, and its estimated value from a reduced form equation is substituted for the observed values in estimating the migration equations. The estimation of CWAGE is described in the next chapter.

coefficient is r_3 in equation 10.) Since the coefficient is positive, we must interpret it as not significant despite the high t value.

GHEPP, another rate of return variable in the place of origin, has both a positive coefficient and a low t value. It clearly is not significant.

Table 18. State-to-City Migration, 1960 to 1970: The Basic Model

Variable	Log		Linear	
	Regression coefficient	t-value	Regression coefficient	t-value
CONSTANT	-1.826	-6.34***	-.865	-3.78***
CWAGE60	.745	4.07***	.286	6.96***
AGINC60	.735	5.87***	.059	0.59
GEDPP60	.155	2.92	.130	1.90
GHEPP69	.012	0.35	.018	0.94
CLAFO60	.086	2.92***	.080	2.41**
SLAFO60	.478	12.97***	.154	3.64***
SCDIST	-.546	-10.10***	-.153	-7.41***
CLMPR60	-.040	-1.98**	.045	0.65
PPYNG60	1.377	2.34***	1.00	3.05***
PPLIT60	.536	3.94***	.071	0.88
MIGST60	.614	24.74***	.192	36.12***
	R^2 = .86	F = 597.	R^2 = .72	F = 259.
n	1,081		1,081	

The "information" variables (CLAFO, SLAFO, SCDIST, and MIGST) are all highly significant and of the expected sign, and the two "response propensity" variables (PPYNG and PPLIT) are significantly positive, as expected.

The logarithmic specification has high explanatory power with R^2=.86 and is highly significant (F=597).

In contrast, the linear specification has less explanatory power with R^2=.72 and F=259. Although nearly all variables save CLMPR are shown to be of the same sign as in the logarithmic specification, several additional variables fail the test of significance. However, MIGST has a substantially higher t-value, suggesting that its effect is linear. We conclude on balance that the logarithmic form is the better specification of the migration relation.

With regard to the information effect of MIGST and CLAFO, several experiments were attempted. First, CLAFO was deleted from the regression to determine whether its information was essentially a reflection of its correlation with MIGST (r=.59). The results supported this possibility, since its deletion had no effect on the equation's explanatory power (R_2 remained at .86), and the F-ratio increased.

We indicated earlier the possibility that the affect of prior i-to-j migration is better specified in terms of distributed lags than in terms of migrant stock (see equations 8 and 10). To test this possibility, the model was run with lagged migration SCMIG50 substituted in the regression in place of MIGST. Although the R^2 for the equation remained unaffected, the t-value for SCMIG50 was somewhat lower than that for MIGST and the

F-ratio for the equation was likewise lower (F=416).[44]

From the standpoint of developing a policy in Mexico to limit migration to cities, the results shown in table 18 are not encouraging, to put it mildly. Some of the most important variables--size of state and city labor forces, distance, age distribution of the population and migrant stock--would be unresponsive to policy except over the very long run. Of those more readily affected by policy--city wage and agricultural income, government expenditures on health and education, urban labor market pressure and literacy--the only socially acceptable policies would increase rather than diminish migration to the cities.[45]

The results in table 18 thus suggest strongly that finding a socially acceptable way to limit migration to cities in Mexico may be very difficult. Subsequent analysis indicates that the problem may not be as intractable as table 18 suggests, although in the end we conclude that the policy instruments by which the Mexican government might affect migration would in fact have only modest impact. This argument is developed in the rest of this chapter and in those that follow.

First Variation on the Basic Migration Model

As indicated in the discussion of table 9 (chapter III), there are

---

[44]As noted earlier, the use of a distributed lag model was explored at some length, but the results were unsatisfactory. An account of this work is in the appendix to this chapter.

[45]Conceivably the social costs of migration to cities might become so high that the Mexican government would seek to control it by depressing urban wages and agricultural income, increasing urban unemployment and reducing expenditures in the states for health and education, but we believe this is so unlikely as to be disregarded.

wide variations among Mexico's 31 states with respect to the percentage of the population living in urban areas. It is plausible to believe that migration behavior might be different for more urban populations than for more rural populations. To explore this idea we computed regressions for the two groups of rural and urban states for which data are shown in table 9. The specification of the regressions is the same as in the logarithmic version of the basic migration model. The results are shown in table 19. They reveal some interesting differences between rural and urban states with respect to migration. Migrants from urban states were strongly attracted by city wages while the attraction for rural state migrants was negligible. Agricultural income was a much stronger factor in inducing migration from urban states than from rural states. (In both cases we interpret the positive coefficient of AGINC to mean that the "ability to pay moving costs" and "propensity to move" components of the coefficients outweigh the rate of return component.) Migrants from rural states were significantly discouraged by the amount of pressure on urban labor markets (CLMPR), while the effect of this factor on migrants from urban states, while negative, was insignificant. Differences with respect to the information variables (CLAFO, SLAFO, SCDIST and MIGST) are mixed. For CLAFO and SLAFO the differences are small, but distance (SCDIST) was a considerably stronger deterrent to migrants from urban states than to those from rural states, while migrant stock was more strongly positive with rural than with urban state migrants.

Responses to government expenditures on education and health (GEDPP and GHEPP) were similar in the two sets of states, but the socio-demographic "response prospensity" variables (PPYNG--percent of the state population age 10-29 and PPLIT--percent of the state population age 6 or more literate) had a strong positive effect on rural state migrants but were of negligible importance in urban states.

Table 19. State-to-City Migration, 1960 to 1970: First Variation on the Basic Model

Variable	Urban states		Rural states	
	Regression coefficient	t-value	Regression coefficient	t-value
CONSTANT	-2.016	-3.40***	-1.615	-2.80***
CWAGE60	1.364	5.22***	.077	0.25
AGINC60	.790	4.69***	.554	2.11**
GEDPP60	.186	2.61	.210	1.98
GHEPP60	.099	1.25	.055	1.21
CLAFO60	.079	1.92***	.105	2.40***
SLAFO60	.573	8.31***	.470	6.61***
SCDIST	-.693	-8.45***	-.373	-4.17***
CLMPR60	-.005	-0.18	-.084	-2.69***
PPYNG60	.990	0.60	2.000	2.35***
PPLIT60	.072	0.18	.817	4.10***
MIGST60	.520	13.94***	.716	19.56***
	R^2=.85	F=275	R^2=.87	F=334

***Indicates significance at the 1 percent level, **at the 5 percent level.

The data are in logarithms.

On balance, the results suggest that migrants from urban states were more sensitive to strictly economic factors and less sensitive to socio-demographic factors than migrants from rural states, while the two groups were about equally sensitive to information and social expenditure factors.[46]

The different responses to economic factors is particularly interesting because any policy to influence migration inevitably would lean heavily on economic incentives. While the results from urban and rural states do not resolve the policy problem revealed by the basic migration model they do suggest that the problem may not be as severe in dealing with migration from rural states as from urban states because the _perverse_ effects of policy, e.g., raising agricultural income, would be weaker in rural states than in urban states.

It is of interest, therefore, to explore further the difference between the two sets of states with respect to responses of migration to economic factors. We do this by calculating the ratio of each city's wage to each state's agricultural income and substituting this variable (SCWAG) for the city wage (CWAGE) in the regressions. The logic is that if migrants from urban states are generally more sensitive to economic factors than those from rural states, they ought to be more responsive to _differences_ between wages in the states and in the cities. The ratios

---

[46]The responses to CLMPR may not fit this generalization since they suggest that urban state migrants were _less_ and rural state migrants _more_ sensitive to this economic factor. The unimportance of this factor to urban state migrants may reflect their higher level of education (see table 9), indicating greater confidence in their ability to get a job in the city for any given amount of labor market pressure.

of city wages to agricultural income per worker pick up these differences. The analysis and results are presented in the next section.

Second Variation on the Basic Migration Model

Table 20 presents the regression results for urban and rural states. They are similar to those shown in table 19. However, the urban-rural

Table 20. State-to-City Migration, 1960 to 1970: Second Variation on the Basic Model

	Urban states		Rural states	
Variable	Regression coefficient	t-value	Regression coefficient	t-value
CONSTANT	-3.43	-4.96***	-1.70	-2.17**
SCWAG60	1.20	5.38***	0.07	0.28
AGINC60	2.25	7.28***	0.65	1.33
GEDPP60	0.18	2.49	0.21	1.99
GHEPP60	0.09	1.11	0.05	1.21
CLAFO60	0.08	1.88*	0.10	2.38***
SLAFO60	0.56	8.33***	0.47	6.62***
SCDIST	-0.71	-8.56***	-0.38	-4.18***
CLMPR60	-0.00	-0.08	-0.84	-2.66***
PPYNG60	1.36	0.83	2.00	2.36***
PPLIT60	0.12	0.29	0.82	4.10***
MIGST60	0.52	13.84***	.72	19.56***
	R^2=.85	F=276.	R^2=.87	F=334.
n	552		529	

***indicates significance at the 1 percent level, **at the 5 percent level, *at the 10 percent level.

The date are in logarithms.

wage ratio (SCWAG60)[47] yields a slightly higher t-value than did CWAGE60 in the regression for urban states in table 19, and AGINC60 for urban states has a substantially higher t-value than it had in table 19. In the regression for rural states, neither SCWAG60 nor AGINC60 is significant. Evidently, the migrating populations from rural states did not make explicit comparisons of urban-rural wage differentials in selecting a designation city, while urban state migrants were very sensitive to these differentials.

The results in table 20 are consistent with the hypothesis that urban state migrants are more sensitive to economic factors--particularly wage differentials--than rural state migrants. They are particularly interesting, however, because they indicate that the "perverse" relationship between agricultural income and migration may be weak to the point of insignificance in the rural states, after the effect of the urban-agricultural income differential is taken into account. Nonetheless, from a policy standpoint, the positive relationship in both sets of states between migration and agricultural income remains troublesome.

In thinking about this problem we recalled that in the distributed lag specification of the migration model, described in the appendix to this chapter, the relation between migration and the <u>growth</u> of agricultural income was negative and highly significant (see table 23 in the appendix). On reflection, this suggested that expectations about future agricultural income may be an important factor in migration decisions; the more positive

---

[47]SCWAG60 is calculated as CWAGE60÷AGINC60. CWAGE60 is calculated following the two-stage least squares procedures, as shown in the next chapter.

the expectations the weaker the incentive to migrate. We decided to explore this hypothesis, maintaining the distinction between urban and rural states.

Third Variation on the Basic Migration Model

In this variation the percentage growth in state agricultural income per worker from 1950 to 1960 (GAGIN) was included in the model specification. Since we had data for only 21 states in 1950, the model was run for only those states. The results are shown in table 21.

Table 21. State-to-City Migration, 1960 to 1970: Third Variation on the Basic Model

	Urban states		Rural states	
Variable	Regression coefficient	t-value	Regression coefficient	t-value
CONSTANT	-1.99	-1.90**	-2.52	-3.14***
CWAGE60	2.11	5.34***	0.67	1.80**
AGINC60	0.79	0.85	-0.57	-1.65**
GAGIN50	-0.04	-1.52*	-0.02	-3.04***
GEDPP60	-0.41	-1.06	-0.10	-0.77
GHEPP60	-0.26	-0.64	-0.06	-0.71
CLAFO60	-0.01	-0.14	0.04	0.82
SLAFO60	0.23	0.91	0.30	2.68***
SCDIST	-1.05	-8.02***	-0.61	-5.54***
CLMPR60	0.02	0.42	-0.05	-1.42*
PPYNG60	0.97	3.27***	7.15	4.20***
PPLIT60	0.24	0.40	0.40	1.57*
MIGST60	0.46	8.92***	0.67	15.37***
	R^2=.87	F=168	R^2=.87	F=229
n	313		425	

***indicates significance at the 1 percent leve, **at the 5 percent level, *at the 10 percent level.
The data are in logarithms.

Except for a minor difference for the urban states, the R^2s in this specification are the same as in the others. The F statistic is lower in this specification than in the others, although still highly significant. The different number of states included may account for some of the differences in F values for this specification, as well as for some of the differences in coefficients.

The results with this specification of the regression model support the hypothesis that migration from 1960 to 1970 was negatively associated with the expected growth of agricultural income, and that those expectations were based on the growth of agricultural income in the 1950s. The hypothesis is most strongly supported in the rural states. The support is less strong in the urban states, where the coefficient for GAGIN is significant at the 6.5 percent level in a one-tail test (t=1.52). The <u>level</u> of agricultural income in the urban states, while still positively related to migration, is not significant. In the rural states the coefficient for the level of agricultural income now is negative and significant at the 5 percent level in a one-tail test.

Also of interest is that two other policy variables, federal expendituress in the states on education (GEDPP) and health (GHEPP), now have the expected negative signs in both sets of states, although neither is significant.

The third variation of the basic migration model supports the argument that policies to increase state agricultural income would be expected to have some modest effect in reducing migration to cities. The expected effects would be modest for two reasons. First, the growth of agricultural income has a small effect in reducing migration. Table 21 indicates that among the rural states a 10 percent difference in growth of agricultural income was associated with a .2 percent difference in migration. The comparable figure for urban states was .4 percent.

Second, the growth of agricultural income must be sustained long enough to stimulate expectations of continued growth. Because of the nature of the data, our analysis shows that migration in one decade was a function of the growth of agricultural income in the previous decade. If annual data were available, they might show that a shorter period would suffice to create positive expectations about income growth. Nevertheless, policies to increase agricultural income may take several years to implement (e.g., construction of irrigation systems), and several more years may be necessary to convince farmers of the effectiveness of the policies. They may require some "convincing," particularly if stochastic events (such as the weather) obscure the effects of agricultural income policies.

Despite these limitations, our analysis indicates that among the factors specified as influencing migration, the growth of agricultural income is the only one which might give some leverage in reducing migration. In the last chapter we explored various factors affecting agricultural income per worker, and found several (see table 13). Among them, the ones most readily affected by policy are number of tractors per hectare (TRCPH), irrigated land per agricultural worker (IRRPL), literacy (PPLIT), and percent of arable land in _ejidos_ (PABEJ). However, it is possible that higher values of the first three of these variables also would induce more migration, thus offsetting the effect of growing agricultural income on migration. (We hypothesize a negative relation between _ejidos_ and migration). In fact, table 21 indicates that increased literacy would stimulate migration from both rural and urban states. This may also be true of tractors since under many circumstances tractors substitute for labor, thus pushing workers off the land and possibly inducing them to migrate.

Table 13 indicates a negative, but not significant, relationship between _ejidos_ and agricultural income. Nonetheless, the Mexican government has long had and continues to have a policy of encouraging the spread of the _ejido_ system of farming as a way of meeting the "land hunger" of Mexican peasants. It is important, therefore, to consider the relation between _ejidos_ and migration, as well as between migration and tractors and irrigation, respectively. We do this in the next section.

Fourth Variation on the Basic Migration Model

Because we wanted to include the growth of agricultural income in the model, this variation is run on the 21 states for which data were available for both 1950 and 1960. We hypothesize that the relationship between _ejidos_ and migration is negative, but that the relationship between migration and tractors and migration and irrigation may go either way. In this instance we did not use the distinction between urban and rural states. The results are shown in table 22. Of particular importance is that the relation of the growth of agricultural income to migration is negative and significant, both when the level of agricultural income is included in the specification and when it is not. The results with respect to tractors, irrigation and _ejidos_ also are encouraging. They indicate that states with more tractors per hectare had _less_, not more migraion, as might be expected, and that states with relatively large amounts of land in _ejidos_ also had less migration. The results for irrigation do not indicate

Table 22. State-to-City Migration, 1960 to 1970: Fourth Variation on the Basic Model

Variable	AGINC & GAGIN		GAGIN only	
	Regression coefficient	t-value	Regression coefficient	t-value
CONSTANT	-3.387	-6.92***	-3.433	-7.00***
CWAGE60	1.449	9.04***	1.413	8.90***
AGINC60	.479	2.15**	-	-
GAGIN50	- .012	-2.82***	- .013	- 3.11**
CLAFO60	- .003	-0.08	- .002	- 0.05
SLAFO60	.435	8.34***	.392	8.18***
SCDIST	- .833	-13.15***	- .812	-13.05***
PPYNG60	5.240	3.83***	6.516	5.27***
PPLIT60	.721	3.68***	.830	4.37***
MIGST60	.571	22.03***	.577	22.43***
PABEJ60	- .047	- 0.61	- .111	- 1.56*
TRCPH60	- .008	- 2.12**	- .005	- 1.35
IRRPL60	- .002	- 0.11	.007	0.56
	R^2=.87	F=388	R^2=.86	F=422
n	737		737	

***Indicates significance at the 1 percent level, **at the 5 percent level, *at the 10 percent level.

The data are in logarithms.

a significant relationship, either positive or negative, between irrigation and migration.

The results do not support the hypothesis that tractors displace labor, thus encouraging migration. This statement must be carefully considered, however. The hypothesis that tractors displace labor assumes implicitly that all other things remain the same, including the quantity of other inputs. In Mexico, however, states which have more tractors generally are agriculturally more advanced states, meaning that they employ more of other kinds of inputs as well as tractors, and also produce more per unit of land. In particular, there is a correlation between tractors per hectare and percent of arable land irrigated (r = .67). Irrigated land typically yields more per hectare than unirrigated land, both because of better water control and because irrigation usually involves more complementary inputs, such as fertilizer, as noted at the end of the last chapter. In addition, irrigation may permit the harvest of more than one crop per year. The higher yields with irrigation generate a greater demand for labor for planting, cultivating, and harvesting, even where tractors are used. In addition, the maintenance of irrigation systems creates demands for labor not needed in dryland farming.[48]

Next Steps

The fourth variation of the basic migration model suggests that policies to stimulate the growth of agricultural income would have some

---

[48]The relation between the growth of irrigation and the demand for tractors, labor, and other inputs is highlighted in the discussion of Sonora in chapter 7.

modest effect in reducing migration to cities, and that if these policies operate through increasing the number of tractors and the amount of irrigated land, they would not offset these effects by displacing labor, thus stimulating migration. On the contrary, the transformation of agriculture represented by more tractors and irrigation may directly restrain migration, although these effects would be slight.

We also know from chapter III, however, that agricultural income, migration, and urban wages are linked in an interdependent system. A complete account of policy alternatives for affecting migration must rest on anslysis of that system. In the last chapter we discussed determinants of agricultural income in the system and this chapter has focused on migration. We must now turn to analysis of determinants of the urban wage. That is the business of the next chapter.

APPENDIX TO CHAPTER 4

The Distributed Lag Model

Assuming that the strength of the information effect of prior migration on current migration declines at a constant rate λ per unit time, then (from equation (10)) the n prior migration elasticity coefficients can be written as:

$$k_{4+\tau} = \lambda^{\tau} k_4, \ \tau = 1,\ldots,n,$$

and

$$0 < \lambda < 1 \ ,$$

so that equation 10 can be written as

$$M_{ijt} = aw_{it}^{r_1} \ldots S_{it}^{p_3} M_{ijt-1}^{\lambda k_4} M_{ijt-2}^{\lambda^2 k_4} \ldots M_{ijt-n}^{\lambda^n k_4} \qquad (10')$$

Estimation of (10') would confront two serious problems. First, we would need time series for migration in previous periods, and we do not have them. Second, even if we had the necessary data, there likely would be high correlation in the M_{ij}'s from one period to the next, making it impossible to obtain unbiased estimates of the parameters for each period.

These problems can be avoided by writing equation (10') for migration in the preceding period, and raising it to the λ power.

$$M_{ijt-1}^{\lambda} = a^{\lambda} w_{it-1}^{\lambda(r_1+p_1)} \ldots S_{it-1}^{\lambda p_3} M_{ijt-2}^{\lambda^2 k_4} \ldots M_{ijt-n-1}^{\lambda^{n+1} k_4} \qquad (10'')$$

and then dividing (10') by (10'')

$$\frac{M_{ijt}}{M_{ijt-1}^{\lambda}} = a^{1-\lambda} \frac{W_{it}^{r_1+p_1}}{W_{it-1}^{(r_1+p_1)}} \ldots \frac{S_{it}^{P_3}}{S_{it-1}^{\lambda P_3}} M_{ijt-1}^{\lambda k_4} M_{ijt-n-1}^{-\lambda^{n+1} k_4} \qquad (10''')$$

Notice that all the lagged values of M_{ij} from t-2 through t-n have cancelled out, leaving only $M_{ijt-1}^{\lambda k_4}$ and $M_{ijt-n-1}^{-\lambda^{n+1} k_4}$. However, since λ is a number between zero and one, the value of λ^{n+1} approaches zero the more rapidly as n is large and λ is small. If λ^{n+1} is approximately zero, the last term of this equation approximates unity.[49] Restating equation (10''') in terms of M_{ijt} would then yield

$$M_{ijt} = a^{1-\lambda} \frac{W_{it}^{r_1+p_1}}{W_{it-1}^{\lambda(r_1+p_1)}} \cdots \frac{S_{it}^{p_3}}{S_{it-1}^{p_3}} M_{ijt-1}^{\lambda(k_4+1)} \tag{11}$$

---

[49] If, for example, the elasticity of migration with respect to lagged migration should decrease by as much as 10 percent in one year, then the annual rate of λ is .9; a migrant 50 years ago would have an elasticity upon current migrants of .005 $k_4 \cong 0$, as shown below.

one-year % decrease in k_4	Equivalent value of λ: one-year rate	ten-year rate	Value of λ^n for 50-year lag
10%	.90	.35	.005
	.93	.50	.031
5%	.95	.60	.078

While the 10% per year decrease, intuitively, appears unrealistically rapid, the 5% annual decrease seems more acceptable. But the associated value of λ^n=7.8% conflicts with the assumption that λ^n=0. A one-year decrease in k_4 of 6-2/3% seems intuitively acceptable and, with a ten year λ rate of .5, yields a λ^n that is not too high for the form shown in equation (10). The actual value of λ is, however, an empirical question.

The estimation of this regression requires, instead of the time series data for migration, only the preceding period's migration (M_{ijt-1}) and the values of all other independent variables also lagged one time period.

However, two estimation problems remain. The first is in the computation of the lag term λ, and its separation from the lagged migration elasticity k_4. This can be resolved if the coefficients of all other terms are unbiased, and if the lag structure is correctly specified. In this case, the regression coefficient for the 1950 value of any variable, excepting M_{ijt-1} and SCDIST, can be divided by the coefficient for its 1960 value, yielding an estimate of λ. This result can then be used to find the value of k_4.

But here, the second estimation problem arises. Table 23 shows that the intercorrelation between the 1950 and the 1960 values of a given variable is frequently prohibitively high (as in the case of CLAFO, SLAFO, and PPLIT). In consequence, the regression coefficients for the variables will be biased, and an unbiased value of λ cannot be calculated. When this regression was run, it provided a slight improvement in R^2 (.88) and a sharp decline in the F-ratio (290) due to the increased number of independent variables. However, using the 1950 and 1960 coefficient pairs for the different independent variables to calculate λ gave estimates ranging from $\lambda = .53$ up to $\lambda = 1.23$, the latter value being unacceptable since λ cannot exceed one.[50]

---

[50] These calculations were made with variables the coefficients of which were significant at not less than the 10 percent level in both 1950 and 1960.

To avoid the collinearity problem, two additional techniques were attemped. First, a ridge regression estimation was made. But this yielded nonsensical values of the coefficients, and these results were rejected.

Table 23. Simple Correlations Between 1950 and 1960 Values of Nine Correlates of Migration

Variable	r	Variable	f
SCMIG60,50	.86	CLMPR60,50	.57
CWAGE60,50	.86	PRYNG60,50	.47
AGINC60,50	.83	PPLIT60,50	.99
CLAFO60,50	∿ 1.00	EDTPP60,50	.93
SLAFO60,50	.99		

Note: The calculations are for the 21 states for which both 1950 and 1960 data were available.

An alternate approach involved the calculation of the ratio of the 1960 value of each variable to its 1950 value with the 1950 value first raised to an arbitrarily set λ power. This ratio was then incoporated in the regression in place of the two 1950 and 1960 intercorrelated value of each variable. The resulting equivalent specification of equation (11) was then estimated

$$M_{ijt} = a^{1-\lambda}\left(\frac{w_{it}}{w^{\lambda}_{it-1}}\right)^{r_1+p_1} \cdots \left(\frac{S_{it}}{S^{\lambda}_{it-1}}\right)^{P_3} (M^{\lambda}_{ijt-1})^{k_4+1} \qquad (11')$$

The results, with λ=.5, are shown in table 24.[51] By parameter testing values of $0<\lambda<1$, with the value of λ increased by .1 at each test, the value of λ yielding the highest R^2 could be identified.

The results provided by this experiment were inconclusive. R^2s varied negligibly, falling from .863 at λ=.1 to .855 at λ=.9. Further, the coefficients and their degree of significance proved unstable. One surprising result, worthy of mention, was that the sign of AGINC60, with AGINC50 held constant, was shown as significantly negative for values of λ ranging from .3 to .9. Also of interest is that the t-value of SCMIG50 exceeded that yielded in the log regression in table 9 by the variable MIGST.

---

[51] It should be noted that, as before, the two-stage least squares procedure was used for estimating the values of CWAGE60 used in deriving the results shown in table 23. Because 1950 values of GEDPP and GHEPP were unavailable, the variable EDTPP (state primary school teachers per capita) was used instead.

Thus, the distributed lag migration specification shown in equation (11') is promising, but because of the unstable behavior of the variables as λ varies, and the inconclusive interpretation of criteria for determining the value of λ, this specification was not further used.

Table 24. Distributed Lag Specification of the Log Form of the Migration Regression, λ=.5

Variable 1960 / Variable 1950	Regression coefficient	t-value
CONSTANT	.445	0.64
CWAGE	1.190	6.37***
AGINC	- .780	- 3.74***
EDTPP	.201	0.60
CLAFO	0.202	2.74***
SLAFO	.717	7.66***
SCDIST	- 1.099	- 12.41***
CLMPR	0.065	0.36
PPYNG	2.479	1.79*
PPLIT	1.697	4.38***
SCMIG50	1.301	29.38***
	R^2 = .86	F = 452
n	738	

Chapter 5

DETERMINANTS OF THE URBAN WAGE

Introduction

This chapter is concerned with the demand for labor in large Mexican cities. The reason for this concern is that policies that affect urban labor markets will impact urban wages and this, as we have seen, will affect migration to cities.

A policy that is intended to reduce rural to urban migration, say by increasing rural income, may, by limiting the supply of labor in urban areas, result in higher urban wages than would otherwise be the case. The stimulus higher wages would give to rural migration may partly offset the initial impact of the migration policy in reducing migration. Furthermore, the policy to raise rural income, if it succeeds, will also increase rural demands for goods and services in nearby cities. The derived impact upon labor demand in these cities will raise urban wages, thus causing an increase in migration to the cities. Although this effect may or may not significantly neutralize rural migration policies in regions receiving public investments, it would affect rural migration rates in regions not receiving such investments, and thereby would have implications for the effectiveness of a national migration policy.

Specification

An aggregate labor demand function can be derived from an aggregation of neo-classical production functions. Sectoral output Y depends upon sectoral employments of labor L and capital K

$$Y = F(L,K) \tag{5.1}$$

where marginal products are assumed continuous and decreasing, and problems of aggregation are assumed insignificant.

If the sectoral production functions of an urban economy, and their aggregate, provide an appropriate specification for a given urban context, then according to neo-classical wage theory, the average wage level W_{ut} in a time period can be stated in terms of the marginal physical product of urban labor, and an average urban commodity price level P_{ut} at that time,

$$W_{ut} = P_{ut} \, (\partial Y/\partial L)_{ut} \tag{5.2}$$

where marginal physical product decreases as L increases.

With capital held constant, equation 5.2 indicates that if the supply of labor available to urban entrepreneurs were diminished due, perhaps, to rural migration policy, then urban wages would rise. It is through this wage effect that we can consider the possibility of induced migration to urban areas that might reverse the initial migration effects of a program to raise rural productivity and income.

In order to consider the additional possibility that the rural productivity program might stimulate local urban output and thereby attract additional urban-bound migrants, it is necessary to consider the dependence of the urban wage level upon urban commodity market behavior.

It is to be emphasized at this point that our intent in this chapter is not the estimation of the urban labor demand function as such, but only of those components of the function that reflect interdependence with the rural sector.

In what follows, we first explain the logic underlying our urban labor demand functions, then the specification actually used, and finally an assessment of the expected signs of the coefficients in light of the underlying logic.

A simple formulation of a commodity price adjustment mechanism in economic dynamics[52] states that the growth in the price level over the level of the preceeding period depends upon the difference between commodity demand (D) and supply (Q) in the preceeding period

$$\frac{Pt}{P_{t-1}} = f(D_{t-1} \div Q_{t-1}). \tag{5.3}$$

Lagged urban commodity demand in this study is assumed to depend in part upon demand in the rural hinterland of the city. If rural demand is a function of the number of rural people employed (L_{at-1}) and of the level of rural income per worker $(Y/L)_{at-1}$, then demand for urban commodities can be written

$$D_{ut-1} = f(L_{at-1}, Y/L_{at-1}, S_{aut-1}, \phi) \tag{5.4}$$

where S_{aut-1} is the share of rural demand that goes to local urban markets, and the term ϕ reflects other determinants of urban demand such as exports to other regions and local consumer demand.

---

[52]R. D. G. Allen, Macro-Economic Theory: A Mathematical Treatment (New York: St. Martins Press, 1968) chapter 5.

Lagged urban commodity supply, assuming neo-classical production functions, can be stated in terms of the lagged urban labor supply and urban capital-labor ratio, the latter determining the marginal production of labor

$$Q_{ut-1} = L_{ut-1} f(K/L)_{ut-1}. \tag{5.5}$$

Although the full spectrum of commodity market effects upon urban labor demand is far more complex than those contained in equations 5.3 through 5.5, the effects of local rural demand are reflected therein.

These determinants of the urban wage are summarized, from equations 5.2 through 5.5, as

$$W_{ut} = f\,(L_{ut},\ L_{ut-1},\ K_{ut},\ K/L_{ut-1},\ L_{at},\ Y/L_{at-1},$$

$$S_{aut-1},\ P_{ut-1},\ \phi_{ut}). \tag{5.6}$$

From the relations underlying this urban wage equation, we derive that current urban wages are lower the larger the level of current or lagged labor supply because: (1) more current labor means lower K/L and hence lower marginal productivity of labor; (2) more lagged labor means more current commodity supply, hence lower commodity prices, hence lower wages. Likewise, a larger current capital supply will increase current labor productivity and therefore wages, while additional capital per worker in the preceeding period means greater commodity abundance which puts downward pressure on current prices, and, thereby, wages. All remaining variables in (5.6) are associated with lagged commodity demand, and are positively associated with the current wage level.

The data requirements and collinearity problems contained in equation (5.6) make its estimation impractical. Nevertheless, our analysis requires the estimation of the impact of the current migration component of L_{ut} and of lagged rural commodity demand upon the urban wage, with additional variables helping to identify the model. The relations in equation (5.6) serve as the basis for interpreting the empirical results.

Data were not available for lagged prices (P_{ut-1}) and non-rural determinants of urban commodity demand (ϕ), nor for the "current" (i.e., 1965) urban capital-labor ratio $(K/L)_u$. However, capital stock and employment by economic sector are available at the national level. These data were used to calculate a capital-labor ratio for each of the 36 cities in 1960 by assuming that in each city the ratios for broad economic sectors (e.g., manufacturing, services) were the same as in the nation. The aggregate ratios differ among cities, therefore, in accordance with differences among cities in the relative importance of the various sectors, measured by employment.

The "current" urban wage is average wages in service industries in 1965. Measures of labor available from the decennial census and computed capital-labor ratios for 1960 served as measures for the lagged time period t-1.

In addition to these data problems, estimation of (5.6) also must deal with the possibility of identification bias in the estimation of the coefficient for current urban labor force (L_{ut}) because of interdependence between L_{ut} and the current urban wage. Since the migration component of current urban labor force is dependent upon the urban wage, as shown in chapter 4, but is also a determinant of urban wage, the urban wage equation should be estimated simultaneously with the migration equation of chapter 4.

Therefore, urban in-migration (M_{ij}) was included in the urban wage specification as a component of the current labor supply (L_{ut}). Natural increase (N_{jt}) is the other component.

The urban wage equation (5.6) specified for estimation, assuming a multiplicative form, is

$$W_{ujt} = \alpha M_{ijt}^{\beta_1} N_{jt}^{\beta_2} L_{ujt-1}^{\beta_3} K/L_{ujt-1}^{\beta_4} L_{ajt-1}^{\beta_5} Y/L_{ajt-1}^{\beta_6} S_{aujt-1}^{\beta_7} \quad (5.7)$$

The effect on wages of additions to city j labor force due to in-migration (M_{ijt}) from origins i[53] is reflected in the coefficient β_1; the effect due to natural increase (N_{jt}) in the city j is shown by β_2. Since these variables reflect the current rather than lagged labor supply, their impact on wages will result from impacts on productivity; both β_1 and β_2 therefore are expected to be negative.

The effect on wages of base year city j labor supply (L_{ujt-1}) is estimated in the coefficient β_3. Whether due to marginal productivity effects or to commodity supply abundance effects, this coefficient is expected to be negative. The current (i.e., 1965) urban capital-labor ratio, which would have a positive effect on current wages by increasing productivity, has not been specified in (5.7) due to lack of data, as noted above.

---

[53] A more appropriate specification would be total in-migration ($\Sigma_i M_{ij}$) to j, with coefficient β_1. Since β_1 is the average effect of origin-specific in-migration which was included for statistical reasons, it may be interpreted as the effect of the average $\overline{M}_{ij} = \Sigma_i M_{ij}/n$. If all cities j have the same number of origins (n), then β_1 will be a scalar transformation of β_1'.

The effect of lagged city j capital-labor ratio (K/L_{ujt-1}), shown in the parameter β_4, is negative, reflecting relative commodity supply abundance. The estimation of β_4, however, will be biased. This is due to the undoubtedly strong correlation between lagged and current capital-labor ratios, the latter of which was not included in (5.7), but would be positively associated with wages. This, and other issues of estimation, will be described in the next section.

Finally, the coefficients β_5, β_6, and β_7 all reflect the effects on wages of lagged excess urban commodity demand, and all three are expected to be positive.

Estimation Problems

The estimation of the parameter β_1 showing the effect on wages W_j of origin-specific in-migration M_{ij} is subject to identification bias since M_{ij} is the dependent variable in the migration equations of chapter 4, with W_j being an independent variable. Therefore, the estimation of β_1 must, as indicated earlier, make use of a simultaneous equations estimation method.

The estimation of the parameter β_3 showing the effect on current wages W_j of lagged labor force L_{ujt-1} is subject to both specification bias and aggregation bias. Specification bias arises because lagged labor supply is also a term in the identity $L_t \equiv \Sigma_i M_{ij} - \Sigma_j M_{ij} + N_j + L_{t-1}$, and therefore may affect wages both through current marginal productivity and through lagged commodity supply abundance. ($\Sigma_j M_{ij}$ is migration from the city.) Aggregation bias results since both the productivity and supply abundance relations are, properly stated, sectoral rather than aggregate phenomena. The aggregation across sectors therefore introduces the possibility of bias and there is strong reason to suspect that it exists, since the summation of sectoral labor supplies also serves as a measure of urban scale. Large cities may contain more specialized and more technologically advanced industries. If such industries pay higher wages, then β_3 may be positive rather than negative. Therefore, a two-tailed test of significance will be required.

If the magnitude of migration and of natural increase are also correlated with urban scale, then the estimation of the parameters β_1 and β_2 would also be subject to aggregation bias. This possibility is mitigated by dividing both migration and natural increase by the base period urban labor supply. These two independent variables are thereby stated as migration rates (M_{ij}/L_{jt-1}) and natural increase rates (N_{jt}/L_{jt-1}) for estimation purposes, and a one-tailed test may be used to test for the expected negative signs.

Data for current natural increase (1960-1965) were not available. Lagged natural increase (1950-1960) was used as a proxy since the correlation of the two natural increase rates across cities is expected to be high.

The deletion of the current capital-labor ratio may, if correlated with the beginning of period capital-labor ratio, cause specification bias in the estimation of β_4, as indicated earlier. Since current K/L is expected to be positively associated with wages, a two-tailed test of significance is required.

Finally, the deletion of lagged non-agricultural determinants of urban commodity demand ϕ, and of lagged average prices P_{ut-1}, is not expected to introduce bias. But since much of the important "export base effect" upon labor demand is thereby not contained in equation (5.7), the coefficient of determination (R^2) will be substantially reduced. This issue is not, however, of concern for this estimation because our interest in the urban wage equation is to capture the wage and migration interdependencies between urban labor markets and rural labor markets, rather than in modeling all determinants of urban growth.

Definition and Measurement of Variables

With two exceptions, all variables to be included in the analysis were also included in the analyses of the preceding chapters. The two new variables are (1) the urban capital-labor ratio, which was estimated from data obtained from the Bank of Mexico and Censo General de Poblacion, 1960; also (2) the city j share of state J urban labor force, data for which was obtained in the Censo General de Poblacion, 1960.

Dependent Variable

$CWAGE_{jt}$: average urban wage, in pesos, in city j, in 1965.

Independent Variables

$GMGLF_{ijt}$: Total migration to city j from each state i, 1960-1970 divided by city i's labor force in 1960. Represents M_{ijt} in equation (5.7).

$CLMPR_{jt-1}$: city j natural population increase from 1950-1960, divided by city j labor force for 1960. Represents N_{jt} in equation (5.7).

$PCSUR_{jt-1}$: ratio of city j labor force in 1960 to total state j urban labor force in 1960. Represents S_{aujt-1} in equation (5.7). We assume that each city in state j shares in rural demand in the state for urban commodities in proportion to the city's share of total urban labor force in the state.

$RURLF_{jt-1}$: rural labor force in state j in 1960. Represents L_{ajt-1} in equation (5.7).

$AGINC_{jt-1}$: output per worker in agriculture, in pesos, in state j in 1960. Represents Y/L_{ajt-1} in equation (5.7).

$CLAFO_{jt-1}$: the labor force in city j in 1960. Represents L_{ujt-1} in equation (5.7).

$CKLRA_{jt-1}$: city capital-labor ratio in 1960. Represents K/L_{ujt-1} in equation (5.7).

Table 25 shows that the average urban wage (CWAGE) is strongly correlated with average agricultural income (AGINC) of the same state ($r = .71$); also, cities with higher wages tend to have experienced slower past labor force growth due to natural increase ($r = .46$ with CLMPR). There are no instances of severe collinearity.

Hypothesis Tests via Multiple Regression

The first regression equation to be evaluated uses the specification of equation (5.7), expressing migration and natural population increase as rates as explained above. The equation is estimated both linear in the data and with logarithmic transformation to reflect the muliplicative form of equation (5.7).

Table 25. Correlation Matrix of Variables Assumed Linearly Correlated With Urban Wage, 1965.

	CWAGE	PCSUR	RURLF	AGINC	CLAFO	CKLRA	CLMPR	SCMIG	GMGLF
CWAGE65	1.00								
PCSUR60	.20	1.00							
RURLF60	-.29	-.50	1.00						
AGINC60	.71	.02	-.50	1.00					
CLAFO60	.28	.52	-.22	.15	1.00				
CKLRA60	-.23	.10	.26	-.30	-.03	1.00			
CLMPR60	-.46	-.13	.02	-.32	-.21	.34	1.00		
GMGLF60	.14	.13	-.13	.22	-.05	-.17	-.15	.26	1.00

Table 26. Average Urban Wage 1965, Estimated via Linear and Log Regression

	Linear		Log	
Variable	Regression coefficient	t-value	Regression coefficient	t-value
CONSTANT	1,820.	1.45	.159	2.67
PCSUR60	4,140.	7.69*	.022	1.28
RURFL60	.00571	6.66	.061	5.04
AGINC60	1.837	24.32***	.795	20.22***
CLAFO60	.00041	1.12	.141	10.50**
CKLRA60	-3,980.	-1.52	-.096	-0.80
CLMPR60	-4,760.	-6.07	-.676	-10.47**
GMGLF60	-62,400.	-2.80***	-.000	-0.01
R^2/F	.62	172.5	.61	164.3
n	730		730	

Asterisks *, **, and *** indicate significance at the 10, 5, and 1 percent levels, respectively.

The linear form shows that the lagged commodity demand determinants stemming from the state rural sector are all positively associated with urban wages. However, the t-values for these variables are inflated, and must be adjusted downward to reflect the correct number of degrees of freedom (the only t-values that should not be adjusted downward are those for variables containing gross i-to-j migration flows).[54] Thus the urban share variable (PCSUR) is significant at the 10 percent level using a one-tailed test. The rural buying power measure (AGINC), however, is significant at the one percent level.

---

[54]The adjustment formula is that written in the note to table 13, chapter 3. In the present case, n in that formula is 36, the number of cities.

Neither of the two lagged urban commodity supply variables (CLAFO60 and CKLRA60) are signficiant in the linear form.

The two variables reflecting the impact of current labor hire upon productivity (CLMPR and GMGLF) are of the expected negative sign, but only the in-migration rate (GMGLF) is significant, that being at the one percent level.

The R^2 for the equation is .62; the F-statistic is shown as 172.5, but must be adjusted downward for the reason explained above.

The logarithmic regression shows nearly as high explanatory power (R^2 = .61) as the linear form, and the signs of all variables are the same as in the linear form. Both AGINC and CLMPR are significant at at least the 5 percent level and of the correct sign. CLAFO is positive and with a two-tailed test, is significant at the 5 percent level. It indicates that larger cities in Mexico tend to yield higher wages to scale and other structural effects.

Because the calculated values of the capital-labor ratio (CKLRA60) were suspected to be subject to substantial measurement error, and since the measure varied little from one city to another, it was decided to test the effect of excluding this variable from the regression. The results showed practially no effect upon the regression coefficients of either the linear or the logarithmic forms. Further, the levels of significance of the coefficients of the remaining variables were unchanged and R^2 was unaffected.

Two additional tests were made of the nature of the simultaneity of the urban wage equation with the agricultural income equation of chapter III and the migration equation of chapter IV.

The first of these tests considered the possibility that a superior specification might take the wage that entrepreneurs currently bid for urban labor as dependent upon the amount of labor demanded at the end of the preceding period. The amount demanded, in turn, would depend upon the commodity supply-demand gap at the end of the preceding period. (See the discussion above of equations (5.3)-(5.6).) The implication of these assumptions is that the urban labor market is recursively rather than simultaneously determined.

To test this assumption, lagged 1950-60 in-migration divided by city labor force in 1960 (PMGLF50), was substituted for GMGLF in the urban wage equation. The results with the new variable did not support the recursive model. All coefficients and t-values were nearly the same as shown in table 26, both in the linear and the log cases, with two exceptions. In the linear case, the coefficient of CLAFO increased from .041 to .063 and became significant at the 10 percent level. However, the t-value of the revised labor demand variable (PMGLF50) was insignificant at -0.14. In the log case, the regression coefficient for PMGLF50 shows a positive instead of the expected negative sign (b = .008) with a low t-value 1.45. The recursive labor market hypothesis is therefore rejected in favor of the simultaneous structure indicated in table 26.

Finally, the possibility is considered that the urban wage level (W_{ujt}) is determined simultaneously with agricultural income per worker, rather than being unilaterally dependent upon it. A number of analysts, including Crosson and Friedmann[55], have concluded that rural productivity and income is enhanced for agricultural regions in the hinterland of productive urban centers, and that productivity and income in the urban center is simultaneously raised by demand from the hinterland for an increasing range of sophisticated goods and services for use in agricultural production.

To test this hypothesis of simultaneity between urban wages and agricultural income the two-stage least squares procedure was used with the exogenous variables from the rural income equation in table 11, from the migration equation in table 21, and from the linear version of the urban wage equation in table 26 to obtain a fitted value of agricultural income per worker (AGINC). Substituting the fitted value of AGINC into the urban wage equation yielded a regression coefficient for AGINC somewhat lower than that shwon in table 26: the coefficient dropped to 1.671 from 1.836 and its t-value dropped to 18.93 from 24.32. It still was significant, however, at the 1 percent level.

In parallel fashion, the fitted value of the urban wage was incorporated into the rural productivity equation as indicated in chapter 3 in the discussion of table 13. The results indicated that the urban wage does exert a strong positive influence on rural wages. The regression coefficient for CWAGE60 (fitted values) was .243, with a t-value of 15.01, significant at the 1 percent level.

---

[55]Pierre Crosson, "Impact of Irrigation Investments on Regional and Urban Development," in Proceedings of the International Symposium on Water Resources Planning (Mexico City, Secretaria de Recursos Hedraulicos, 1972). John Friedmann, Regional Development Planning (Cambridge, Mass.: The MIT Press, 1966).

It is concluded that urban and rural wages for cities and farms in the same state are interdependent. This finding indicates that public policies dealing with urbanization and rural productivity will have indirect as well as direct effects on the urban wage. These issues, and others, are considered in the next chapter.

Chapter 6

IMPLICATIONS FOR AN URBANIZATION POLICY

Introduction

The results of chapter 4 showed that increasing state agricultural income per worker would diminish state-to-city migration, and chapter 3 identified several factors--tractors, irrigation, and literacy--which positively affect agricultural income per worker. In the last chapter we showed that there is interdependence among agricultural income per worker, the urban wage, and state-to-city migration. With these various relations defined and measured, we are in a position to consider some of their implications for an urbanization policy in Mexico.

It is not our intention to make recommendations for such a policy appropriate for conditions of the late 1970s and the 1980s. To do so would be presumptuous. Moreover, our analysis has an inherent limitation as a guide to policy. In the appendix to chapter 3 we stated that for policy purposes it would be desirable to state relationships between agricultural income and policy instruments, for example investment in irrigation, in terms of rates of change. This is true also of the relationship between agricultural income and migration. We then could make statements of the following form: If irrigated land per man is increased by X percent, then agricultural income per man will increase by λX percent and migration to cities will be reduced by Y percent. Such statements obviously would be useful in formulating urbanization policies.

Our data do not permit us to make statements of this sort. We have established relationships between certain variables, for example irrigation and agricultural income at a point in time, 1960, and between agricultural income and migration between 1960 and 1970. The only exception, and an important one, to these point-in-time relationships is the relationship between the growth of agricultural income between 1950 and 1960 and migration from 1960 to 1970.

The nature of the relationships established permit us to make such statements as the following: States which in 1960 had more irrigated land per farm worker also had higher agricultural income per worker; and rural states which (1) had higher agricultural income per worker in 1960 and (2) higher growth of agricultural income per worker between 1950 and 1960 had less migration from 1960 to 1970. As guides to policy, these statements would be put in the following form: If in the rural states in 1960 there had been X more hectares of irrigated land per farm worker, and other variables had been at their 1960 mean values, then agricultural income per farm worker would have been higher by λ X and migration would have been lower by Y percent. In addition, if in the rural states the growth of agricultural income from 1950 to 1960 had been X percent greater, and all other variables including agricultural income, had been at their 1960 mean values, then migration from 1960 to 1970 would have been less by M percent.

Clearly, statements about what-would-have-been at a point in time if certain things had been done differently are not as useful for policy purposes as statements about what-will-be in the future if certain things are done differently.

Still, the first sort of statement can be useful if the relationships underlying the statements have some stability. We believe this is true of the key relationships we have studied. Enough is known about agricultural production functions, for example, to give confidence that in many parts of Mexico at present and in the future an increase in the amount of irrigated land per farm worker, taking this to represent an entire technology as noted earlier, will result in higher income per farm worker.[56] Similarly, our finding that in the rural states migration was inversely related both to income and to the growth of income taken together is consistent with theories and observations of migration behavior in many places and over time. In this connection, table 17 in chapter IV is of particular interest. It shows that the pattern of correlations between independent variables and migration to cities in Mexico from 1960 to 1970 was very similar to the pattern of correlations between these variables and the migrant stock in the cities in 1960. Since the migrant stock is a function of all past migrations, the similarity between the two patterns of correlations indicates stability over time in the relationships between migration and the determinants of it.

In the rest of this chapter we draw on the analyses of chapters III, IV, and V to make statements about what would happen in Mexico if certain policies were pursued. To give concreteness to the discussion, the statements are frequently cast in quantitative terms. These statements are to be interpreted in the light of the preceding discussion, and the quantitative results taken as only rough indicators. Obviously, we think these statements are useful, but they are not precise predictions of responses to policies, despite the seeming precision of the quantitative results

[56]The potential for feasible increases in irrigation in Mexico is discussed in the text below. For the argument that irrigation can be treated as a proxy for an entire technology, see p. 54 above.

obtained. Rather, the statements are intended to indicate the <u>direction</u> of the responses to policies--whether changes in quantities will be positive or negative--and whether the changes will be sufficiently large to be of interest from a policy standpoint. Thus the statements identify some policies that might be useful. Development of those policies, however, and their application to the current situation in Mexico would require more up-to-date and complete data than those used in this study.

An Appraisal of Policy Instruments

In chapter 4 we concluded that among the various factors affecting migration to cities in Mexico, agricultural income per farm worker was the only one which might be effectively used as a policy instrument for reducing migration. The results of the third variation on the basic migration model (table 21 of chapter 4) showed that the relationship between agricultural income and migration was particularly strong in the more rural states. That is one reason for focusing an agricultural incomes policy on those states. There are other reasons. The rural labor force of the more rural states was larger than in the more urban states, and any given percentage reduction in migration would involve more people if it occurred in rural states rather than in urban states. Moreover, people in the rural states not only were more numerous, they also were poorer than those in the more urban states (see table 9, chapter 3). This would be a reason to concentrate efforts to increase agricultural incomes in the rural states, quite apart from effects on migration.

We emphasize the expansion of irrigation because we believe this offers more promise for increasing agricultural income per farm worker than any of the other ways of doing this that we have considered. For example, an increase of one hectare of irrigated land per farm worker in the rural states in 1960 would have raised that ratio from .053 to 1.053, an increase of 1887 percent (see table 9). Table 13 of chapter III indicates that such a percentage increase in irrigated land per farm worker, assuming that irrigation is substitutable with other inputs, would have generated an increase of 26.4 percent in agricultural income per worker (1887 x .014).[57]

However, another viewpoint suggested in chapter III emphasized the complementarity between irrigation and other inputs. In this case, it is plausible to consider irrigation as a proxy for an entire technology, and agricultural income can be treated as a function of irrigation alone. The results of doing that, shown in table 14 of chapter III, indicated that an increase of one hectare of irrigated land per worker would raise agricultural income per worker by approximately 1200 pesos. In the rural states this would be an increase of 32 percent (see table 9).

Thus the analysis of chapter III indicates that in 1960 an increase of one hectare of irrigated land per farm worker in the rural states would have increased agricultural income per worker by one-quarter to one-third, depending upon which specification of the agricultural income equation seems most appropriate. How do these increases compare with those yielded by other

---

[57]The regression equation in table 13 is for all states, not the more rural ones. We believe, however, that the regression can be used to estimate the effect of expanding irrigation on agricultural income per worker in the rural states. The reason is that the rural states had very little irrigation in 1960 and as a consequence there was no correlation between irrigation per worker in those states and agricultural income per worker (r^2 = .09). The correlation between these two variables for all states, shown in tables 13 and 14 of chapter III, was almost all in the more urban states. (The r^2 for those states was .64.) To estimate the effect of agricultural income per worker of a large-scale expansion of irrigation in the rural states, where in 1960 there was very little, we in effect take as a guide the experience of the urban states, where there was much.

factors affecting agricultural income considered in chapter III: literacy, _ejidos_, and tractors?[58]

From tables 11 and 13 of chapter III it can be calculated that to match the income effect of a one-hectare per worker increase in irrigation in the rural states, literacy in those states would have to increase by between 72 to 89 percent. Since the literacy rate was 52.5 percent in 1960 (see table 9), it would have to be increased to 90 or 99 percent.

We have no way of comparing the cost of doing this with the cost of increasing irrigation by one hectare per person. More to the point, however, is the fact that the direct effect of such increases in literacy in stimulating out-migration would more than offset their indirect effects in reducing migration. This can be inferred from table 21 of chapter IV, which shows that a 72 percent increase in literacy in the rural states would increase migration by 29 percent (72 percent x .4).

This direct stimulus to rural out-migration would be only partially offset by the indirect income effect: the 26 percent increase in agricultural income per worker resulting from a 72 percent increase in literacy would decrease migration from the rural states by only 15 percent (26 percent x -.57, also from table 21). The calculations for the more extreme case show that an 89 percent increase in literacy would directly increase migration from the rural states by 36 percent while the indirect effect by way of agricultural income would decrease migration by only 18 percent.

[58] In chapter III we also considered the relation to agricultural income of the size of the rural labor force (RURLF), number of teachers per capita (EDTPP), the rate of migration (OMRLF), and state urban income (AUWAG). However, RURLF and AUWAG are not considered policy variables and EDTPP was discarded because of high correlation with literacy. OMRLF is the _object_ of policy, not an instrument of policy.

Of course, there are excellent reasons for increasing the investment in literacy in the rural states, not the least because of the effect in raising agricultural income. From the standpoint of reducing migration to cities, however, investment in literacy would be counter-productive.

Table 13 of chapter III indicates that expansion of land in _ejidos_ (PABEJ) would have a slight negative but not insignificant effect on agricultural income per worker. Table 22 of chapter IV indicates that the expansion of _ejidos_ would have some small direct effect in restraining migration. We conclude that expansion of the _ejido_ system of farming, apparently still an important policy objective of the Mexican government, likely would have neutral to slightly favorable effects as an instrument of urbanization policy. By comparison with irrigation it is insignificant in this respect.

From table 13 it can be calculated that to match the agricultural income effect in the rural states of a one hectare per man expansion in irrigation (26 percent or 32 percent), the number of tractors per hectare would have to increase by 2600 percent or 3200 percent. Since the number of tractors per thousand hectares in the rural states was 1.09 in 1960 (table 9), that number would have to rise to 29.4 or 36.0.

As in the case of literacy, we are not in a position to compare the cost of these investments in tractors with the costs of expanding irrigated land per man in the rural states. Comparison with the number of tractors per hectare in the United States, however, suggests that the tractor investments would have made little sense for Mexico. In 1959 there were 28 tractors per hectare of cropland in the United States.[59] Since agricultural

---

[59] U.S. Department of Agriculture, _Agricultural Statistics_, 1973.

labor in the rural states of Mexico was much more abundant relative to capital than in the United States, a policy that would have made the number of tractors per hectare in those states equal to or more than in the United States surely would have been most unwise.

Feasibility of Expanding Irrigation

There would be little point in discussing irrigation as an instrument of urbanization policy if there were no feasible opportunities for expanding irrigation in Mexico. In fact, however, there seem to be many such opportunities. In the National Water Plan (PHN) of 1975, the Mexican government projected 4 million hectares of newly irrigated land between 1975 and 2000. While much of this additional land would be in urban states already equipped with large irrigation projects, such as Sonora, Sinaloa, Baja California, and Tamaulipas, a substantial amount would be in rural states in which there now is relatively little irrigation. For example, in the southeastern region the PNH identifies 1.08 million hectares of land feasible for full-scale irrigation, another 359,000 hectares for full-scale irrigation and drainage development, and another 142,000 hectares where drainage and supplemental irrigation projects appear promising. Most of the states in this region were rural in 1960 (Veracruz, Tabasco, Oaxaca, Chaipas).

Since the 1960s Mexico has implemented its Small Irrigation Program (Programa de Pequena Irrigacion--PPI) designed to supplement its older program of building large irrigation projects. The PPI expanded rapidly in the 1960s and 1970s, and the PNH calls for this to continue. The PPI relies on construction of small dams and storage areas, river diversion schemes, and development of groundwater. The areas irrigated typically run from a few hundred hectares to a few thousand. The program is designed to provide irrigation to farmers scattered around the country in areas where they could not be reached economically with large-scale projects. While we lack detailed information on the present location of these projects or where they will be developed in the future, we are confident that by the nature of the program, many of the projects will be in states which we have classified as rural.

In addition to the building of new irrigation works, the PNH puts much emphasis on improving the efficiency of water use in existing irrigation districts as a way of extending the irrigated area or permitting double-cropping of existing irrigated land. The potential for doing this apparently is large. For example, it is estimated that in the northwest, where the Mexican government for some time has had a program to improve water use efficiency, an increase of 10 percentage points in conveyance and field efficiencies would save enough water to double-crop 200,000-250,000 hectares.[60] This alone is 4 to 5 percent of the presently irrigated area of the entire country.

It may be that Mexico will not achieve the projected increase of 4 million hectares of irrigated land by the end of the century, either because of changes in direction by the government or because on close inspection some of the projects prove unfeasible. That an increase of several million hectares will occur, however, seems likely. We conclude that the potential for expansion of irrigation is sufficiently great to warrant assessing the impact of investments in irrigation as an instrument of urbanization policy.

---

[60]World Bank, "Mexico Economic Report Annex on Agriculture," (Unpublished report, November, 1977).

It is important to emphasize that the investments in physical works required to convey water to the farmer's field must be accompanied by others required to provide him adequate and reliable supplies of inputs complementary to water and facilities for marketing his crop. The PNH recognized that in the past these supplemental investments were not always adequate, e.g., credit was not always available or was too costly, fertilizer sometimes was not available when needed, poor roads exacted high costs of transport, and so on. Plans for future expansion call for better performance in these respects than in the past. If this better performance is achieved, the response of agricultural income to the expansion of irrigation ought to be greater than our analysis shows it to have been in the past.

In considering the relation of irrigation to agricultural income we have nothing to say about whether the increase in income is sufficient to justify the investment in irrigation and complementary infrastructure and inputs. This would require an exercise in cost-benefit analysis which is not part of this study. For our purposes it is sufficient that the Mexican government plans a substantial expansion of irrigation in the 1980s. These plans indicate that in the judgment of the Mexican authorities the economic and other social objectives served by irrigation justify the cost of the proposed expansion. Our purpose is to show the effects of the expansion on migration to cities, thus providing additional information for judging the desirability of the expansion from the social standpoint.

Systematic Urbanization Effects of Irrigation Development

The increase in agricultural income resulting from an expansion of irrigation will reduce migration to the cities, thus reducing also the cities' labor supply. Traditional theory predicts that wages in the cities will rise in response, and thereby attract back some of the potential migrants. In tug-of-war fashion, rural wages rise again as these migrants leave, thus slowing the exodus and triggering another partially offsetting wage response in the destination cities. This process of interregional wage adjustment converges until both wages and migration reach equilibrium.

We use the models developed in the preceding three chapters for limited exploration of three aspects of this process of adjustment to an initial investment in irrigation: (1) the effect on Mexico City's wage and on migration to Mexico City from the target state (i.e., the one where the investment is located) and from other states; (2) the effect on agricultural income and migration from Mexico City's rural hinterland (here taken to be the rural state of Mexico, which surrounds the city); (3) the effect on wages in cities in the target state and on migration to those cities from all other states.

The work of the three previous chapters indicates that an expansion of irrigation in a rural state would reduce migration from that state to <u>all</u> cities, not just Mexico City. However, to focus on all cities simultaneously would have required that we structure our simulator as a general equilibrium model. Because of budgetary constraints we were unable to do this. However, the focus on Mexico City gives the flavor of the interregional adjustments in wages and migration that would occur in response to an expansion of irrigation in a rural state. Moreover, Mexico City, as demonstrated in chapter II, is by far the main attraction to urban bound migrants in Mexico, and a major issue of urban policy is how, if at all, this flow may be slowed.

A flow diagram that describes the systematic urbanization effects of investment in irrigation is shown in diagram 3. This diagram indicates the significant relations obtained in the analyses of chapters 3 through 5, but deletes all variables that would not change in response to the investment. This simplification was obtained by transforming all equations so that the variables would be expressed in percent change terms in the case of logarithmic equations or in first difference terms in the case of linear equations. All variables for which no change would occur thereby have zero values and could be excluded from calculations.

Diagram 3 shows the relations used in the simulation. Public investment in irrigation targeted for a given rural state (block 1) would increase average rural income and, likewise, the growth rate of rural income in that state (blocks 2 and 3). Together, these increases would reduce the rate of out-migration from that state (block 4), causing a reduction in the migration rate to Mexico City (block 5). In consequence, Mexico City's wage level would rise slightly (block 6), thereby attracting a partly offsetting flow of migrants from the various states (blocks 4, 7, 8, and 9). However, Mexico City's small wage change would trigger an increase in the average rural income and growth rate in rural income of adjoining Mexico State (blocks 10 and 11). This would reduce the rate of out-migration from that state to Mexico City.

However, while migration to Mexico City would be reduced by the investment in irrigation, migration to cities in the targeted rural state would increase. This effect would be triggered by an increase in the wages of those cities (block 13) as the initial growth in the rural sector's incomes (block 2) stimulates rural demand for commodities and services supplied by the cities. In consequence, migrants would be attracted to cities in the target state (blocks 4, 7, 8, and 9), subject, however, to equilibrium

Diagram 3. The Urbanization Effects of Rural Public Investment

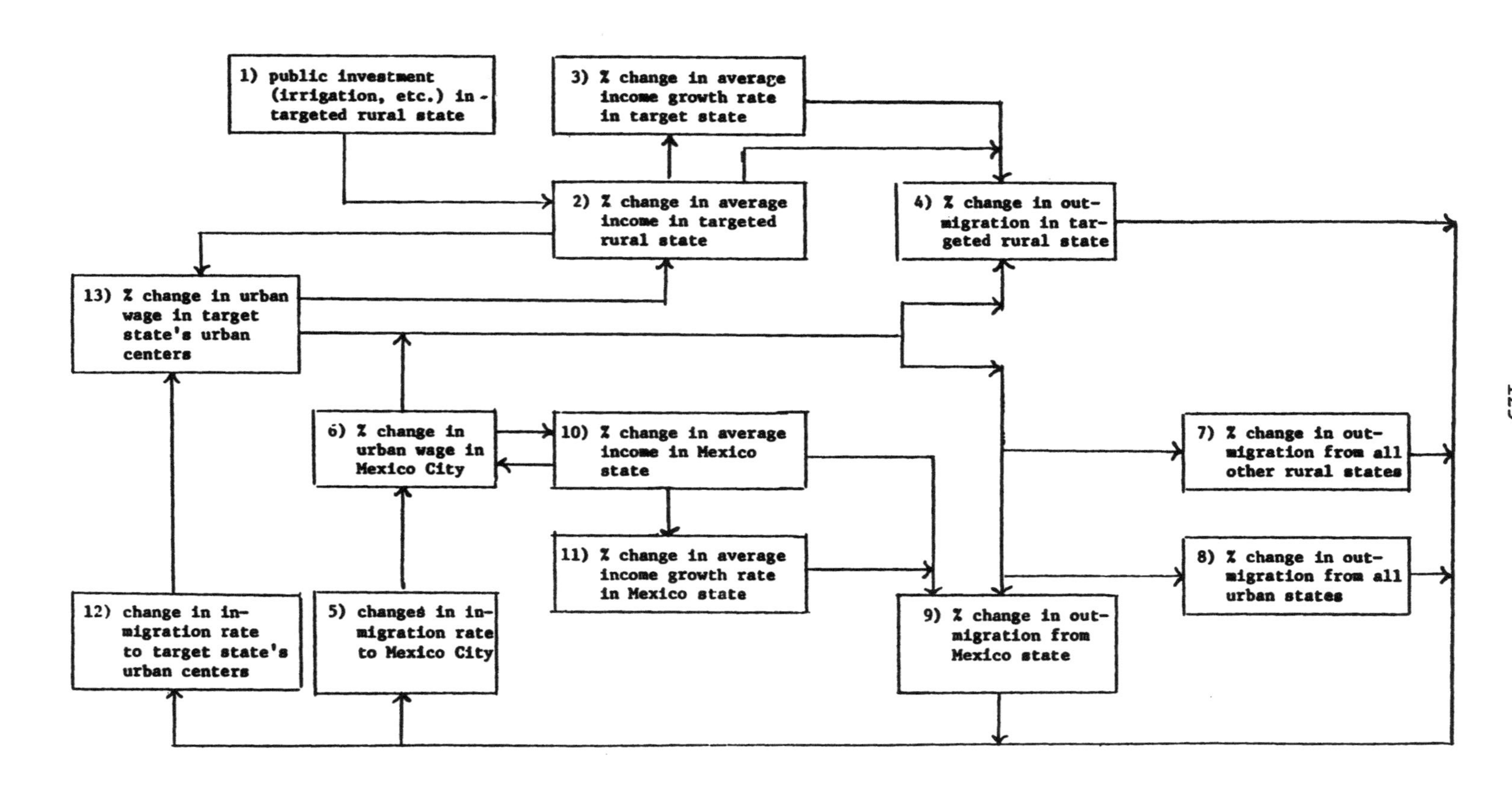

processes as their increased in-migration rates (block 12) cause adjustments to wages (block 13).

Rather than single out a particular target state for these calculations, a composite "state" was used having the attributes similar to the mean values of all rural states in Mexico in 1960. Thus, the observed mean 1960 rural state income per worker of 3741 pesos was used as the measure of income per worker in the "typical" rural state. The values used for the calculations are shown in table 27.

All analyses were carried out to five iterations, at which point convergence was sufficiently complete to make additional iterations unnecessary.

As noted above, an increase of one hectare of irrigated land per worker in a rural state would increase agricultural income per worker by 26 percent or 32 percent, depending upon the specification of the agricultural income equation. To illustrate the interregional effects on income and migration of irrigation development, we take the larger of these two figures.

From Table 21 of chapter IV it can be calculated that a 32 percent increase in agricultural income in a "typical" rural state would reduce migration to cities by 19.2 percent in the decade following the income increase.[61] Our simulation indicates that the interregional adjustment process thus set in motion would reinforce these initial effects in the target state of the expansion of irrigation. The initial reduction in

---

[61]This is obtained by multiplying the percent increase in rural income $\frac{(\Delta AGINC)}{AGINC}$ by the coefficient -.57 and adding to this the results obtained by multiplying the percent increase in the growth rate in rural income $\frac{(\Delta GAGIN)}{GAGIN}$ by the coefficient -.02: -.57 (32%)-.02(60%) = 19.2%.

Table 27. Attributes of the "Typical" Rural State and of Mexico City Used for the Impact Analysis

Variable	Value
AGINC60, "State"	3,741 pesos
AGINC50, "State"	1,764 pesos
CLAFO60, "State"	50,000 workers*
CWAG60, "State"	8,182 pesos
CWAG60, Mexico City	13,608 pesos
CLAFO60, Mexico City	1,749,900 workers
AGINC60, Mexico State	4,035 pesos
Rural "State" to Mexico City MIG	23,388 migrants
Urban "State" to Mexico City MIG	7,768 migrants
Rural State to Local City MIG	1,000 migrants
Mexico State to Mexico City MIG	53,491 migrants

*This variable was approximated; all others are observed mean values.

migration from the state would be slightly offset by wage increases in Mexico City; however, induced expansion in the state's urban areas and the positive effect of this on rural wages would reduce migration from the state far more than higher wages in Mexico City would increase it. Ultimately, the reduction in migration from the state would be 25 percent, not the initial 19.2 percent, and the initial 32 percent increase in average income per farm worker would rise to 44 percent.

The state's large city, assumed to have an initial labor force of 50,000, is calculated to experience an induced in-migration of 4,630. If its labor force participation rate were 40 percent (Mexico City's was 37 percent), then the local city's population would have grown 3.7 percent because of the irrigation program.

Urban wages in Mexico City would rise by only .05 percent, and the impact of this on the city's rural hinterland (the state of Mexico) would be nil. Finally, a net of 5,594 fewer people would migrate to Mexico City, a decline in its decade rate of in-migration of only 1.2 percent.

In sum, an irrigation project in a rural state that would increase irrigation by one hectare per worker would substantially increase rural and within-state urban wages, would bring an increase in the state's large urban center of under 4 percent, and would reduce migration to Mexico City from all Mexican states by 1.2 percent.

Although these impacts indicate a slight shift in the pattern of urbanization toward secondary cities serving rural hinterlands and away from Mexico City, there is no doubt that the benefits generated by the rural public investment largely lie in the direct within-state rural income and migration effects.

Our analysis simulated the effects of an expansion of 218,000 hectares of irrigated land in a "typical" rural state.[62] If comparable investments were made in additional states—say ten—the consequences for the national pattern of urbanization likely would be significant, although we have not systematically worked them out. Our judgment, however, is that in addition to stimulating modest urban population growth in the several states, the migration stream to Mexico City would ge reduced by 12 to 16 percent.

---

[62] The "typical" rural state had an agricultural labor force of 217,990 people in 1960 (see table 9) and we assumed an expansion of one hectare per person.

Chapter 7

IMPACT OF IRRIGATION INVESTMENTS ON AGRICULTURAL AND URBAN DEVELOPMENT IN SONORA

Introduction

The results presented in the last chapter suggest that a large-scale program of public investment in irrigation, or in other projects with comparable effects on agricultural income, could play a role in the implementation of a policy to achieve urban decentralization. The analysis yielding these results, as with many such econometric exercises, was technical and forced to rely on sometimes questionable data and assumptions. We think it useful, therefore, to take an entirely different approach to explore some of the same issues. That is the purpose of this chapter.

In chapter 5 we hypothesized that urban wages and agricultural income per worker in a given state are interdependent, and the econometric analysis of chapter 5 supported the hypothesis. An implication of the hypothesis is that investments to develop agriculture in a state will also indirectly stimulate the growth of urban areas in the state, and the urban growth will have positive feed-back effects on agriculture. In addition, the indirectly induced expansion of urban economic activity in the state will increase the attractiveness to potential migrants of those urban areas relative to Mexico City and other cities. Thus the inter-dependence between agricultural growth and urban growth in a given state played a central role in the disucssion in the preceding chapter of policy implications of the econometric analysis.

In this chapter we explore the relationship between agricultural development and urban growth through analysis of experience in the state of Sonora. The main advantage of this approach is that by concentrating on a single state we can probe the relationship in much greater depth and with more flexibility than is possible with the econometric modeling approach. The main disadvantage is a loss of generality. The experience of Sonora with respect to the expansion of irrigation and consequent agricultural and urban development is not typical of the states of Mexico. In particular, the more rural states where large irrigation investment likely will occur in the future (in the southeast, for example) are more heavily populated and less developed than Sonora was when large public investments in irrigation began there. Also those other states differ from Sonora with respect to climate, soils, proximity to the United States, and other factors relevant to irrigated agriculture.[63]

Despite these differences, the analysis will show that the relationships in Sonora among irrigation, agricultural income, migration, and urban development in the state were generally the same as those established for the country as a whole in the statistical analyses of previous chapters. We believe the experience of Sonora, therefore, supports the conclusions of the last chapter concerning the role of irrigation investments in an urbanization policy.

---

[63]Apparently proximity to the United States contributed something to the generally successful response of agriculture to irrigation in Sonora. There are strong similarities with respect to soils and climate among Sonora, southern Arizona, and southern California. There also was some (unmeasured) flow of capital and managerial expertise from those states to Sonora.

Irrigation and Development in Sonora[64]

The theory of the economic base is useful in analyzing the impact of irrigation on agricultural and urban development in Sonora. There is an extensive literature on this theory. That part which is most relevant here stems primarily from the work of Harold Innis in Canada and of Douglas North in the United States.[65] In this part of the literature, attention is focused on the process by which the development of a region's natural resources provides the impetus to subsequent economic and urban development of the region as a whole. The argument begins with a situation in which for one or more reasons there is increased demand for natural resource-based commodities of the region. This may reflect either a shift in the pattern of demand in favor of the region or productivity improvements in the region's natural resource industries which permit them to increase their penetration of existing markets. Investments in irrigation which increase the productivity of land and other resources employed in agriculture

---

[64] This account draws heavily on Pierre Crosson, "Impact of Irrigation Investments on Regional and Urban Development," Proceedings of International Symposium on Water Resources Planning (Mexico City, Secretariat of Water Resources, December 1975); and "Urban-Rural Relationships in the Northwest Region," chapter 3 in Ronald Cummings, Interbasin Water Transfers: A Case Study of Mexico (Baltimore, Johns Hopkins University Press for Resources for the Future, 1974).

[65] Harold Innis, The Fur Trade in Canada: An Introduction to Canadian Economic History (New Haven, Yale University, Press, 1940); Douglass North, "Location Theory and Regional Economic Development," Journal of Political Economy, vol. 63, no. 3 (1955); and "Agriculture in Regional Economic Growth," Journal of Farm Economics, vol. 41, no. 5 (1959). Some other representative work in this field is by Morgan Thomas, "The Export Base and Development Stage Theories of Regional Economic Growth," Land Economics, vol. 40, no. 4 (1964); and J. C. Stabler, "Exports and Evolution: The Process of Regional Change," Land Economics, vol. 44, no. 1 (1968).

illustrate this latter situation. While in principle the increase in demand can be internal to the region, the analysis typically deals with new or scarcely developed regions where the local market still is very small. Hence emphasis is on the export demand for the region's products.

The growth of demand for regional exports attracts capital and labor into the export industries. While some of these resources may be diverted from other activities within the region, in the cases of interest here most of the additional resources flow in from outside. There is thus a substantial net increase in the quantity of resources employed in the region and in the quantity of production centered on the export industries. The higher level of export activity stimulates growth of the regional economy in four ways by (1) increasing the income and hence the local expenditures of those engaged in export production; (2) increasing demand for locally produced inputs used in the export industries; (3) stimulating the growth of natural resources processing industries (e.g., flour mills to process regionally grown wheat prior to export); (4) increasing the productivity of regional resources, thus permitting local production to substitute for imports or to penetrate external markets from which it had earlier been barred by high production costs. The improvement in productivity may reflect economies of scale achieved as local production rises or the increased skills of labor and management acquired from exposure to a wider range of more advanced activities in the export and export-related industries.

Many of the goods and services demanded by the rising export sector and by people employed in it will be of a type most economically provided in urban area. Wholesale and retail trade, credit facilities, and collection points for transportation of goods into and out of the region are examples of activities for which demand rises sharply as exports grow and which tend strongly to cluster in urban places. Hence the expansion of exports not only stimulates the growth of nonexport production in the region, but also spurs urban development, a crucial element in the economic transformation of the region. Because of the key role they play in initiating and sustaining regional development, export activities are said to constitute the economic base of the region.

Economic base theory provides useful insights into the process of development in Sonora. In 1940 the population of Sonora was 364,176, 5.1 percent of which was located in the state's only urban area, the town of Hermosillo with a population of 18,600. The second town, Ciudad Obregón, had a population of 12,500.[66] Measured by employment the economy of the state was predominantly agricultural. Land in agriculture was about 260,000 hectares of which some 140,000 hectares were irrigated. Virtually all the irrigated land was watered by private wells, or by unregulated stream flow. The total value of agricultural production (not including livestock) was approximately 4 percent of total national agricultural production. The most important crop by far was wheat grown on irrigated land. There were but 444 tractors used in the entire state.[67]

---

[66] Luis Unikel, "El proceso de urbanizacion en Mexico," Demografia y Economia, vol. 2, no. 2 (1968).

[67] Data for state population and labor force are from the 1940 Censo de Poplación, Dirección General de Estadistica, Mexico, D.F. Data on land in agriculture, irrigated land, the value of agricultural production, and the number of tractors are from the 1940 Censo Agrícola Ganadero, Dirección General de Estadistica, Mexico, D.F.

In 1950 agriculture continued to be the dominant occupation in Sonora with more than 54 percent of the state's labor force employed in that sector. From 1940 to 1950 total land in agriculture increased over three times to about 800,000 hectares while irrigated land doubled to 280,000 hectares. Most of the newly irrigated land was from private wells. Although crop production in the nation as a whole increased at the rapid rate of 6.4 percent annually in the 1940s, the share of Sonora in the national total increased slightly. Irrigated wheat continued to be the dominant crop although cotton production moved up from almost nothing in 1940 to a strong second place in 1950. The number of tractors in the state increased almost five times in this period to 2145.

The expansion of the state's agriculture was associated with rapid growth also in both total population and urban population. Total population was 510,000 in 1950, reflecting average annual growth of 3.4 percent, well above the 2.7 percent rate experienced by the country as a whole (see Table 1). Hence there almost surely was net migration to Sonora in the 1940s. Urban population in 1950 was 153,000, located in six towns of which Hermosillo was the largest (43,500 inhabitants) followed by Ciudad Obregón (31,000 inhabitants). The growth rate from 1940 to 1950 for urban population in the state was 23.5 percent annually. Hermosillo's population grew at an annual rate of 8.9 percent while the rate for Ciudad Obregón was 9.5 percent. Clearly net in-migration contributed significantly to the growth of both towns.

It appears that by 1950 most of the opportunities in Sonora for additional expansion of irrigation by private initiative had been exhausted. This can be inferred from the fact that virtually all new irrigated land brought in after 1950 resulted from massive public investments in dams and associated distribution systems. The first major project was a dam begun on the Yaqui River near Ciudad Obregón in 1947 and brought into operation in 1951. A second large irrigation district, created by construction of a dam on the Mayo River, also came into operation in 1951. In 1953 the Mexican government formed a third district to assume control of groundwater used for irrigation in the vicinity of Hermosillo. The projects increased the irrigated area in the state from 280,000 hectares in 1950 to 544,000 in 1960. A few small projects were undertaken in the 1960s but the total irrigated area was not changed significantly. In the late 1960s the three major districts--Yaqui, Costa de Hermosillo, and Mayo, in that order--contained almost 85 percent of the state's irrigated area.[68]

As in the 1940s, the expansion of agriculture in Sonora from 1950 to 1970 was accompanied by rapid population growth, both in the state as a whole, and especially in urban areas. The state's population grew 4 percent annually over the two decades, significantly faster than in the country as a whole (see Table 1), suggesting net in-migration to the state. In the two principal cities, Hermosillo and Ciudad Obregón, population grew at annual average rates, respectively, of 8.3 percent and 8.2 percent in the 1950s and 5.9 percent and 5.3 percent in the 1960s. Clearly net in-migration contributed significantly to each city's population growth in each decade.

---

[68]The figure for 1960 is from the Censo Agrícola Ganadero for that year. All other information is from Caracteristicas de los distritos de riego, Tomo I, 1969, Secretaria de Recursos Hidraulicos (SRH), Mexico, D.F.

The relationship between expansion of irrigation and urban growth in Sonora can best be explored by concentrating on the experience of Ciudad Obregon and its agricultural hinterland, the Rio Yaqui irrigation district, and on Hermosillo and its hinterland, the Costa de Hermosillo irrigation district.

With the construction of the two irrigation districts, agricultural production in the hinterland of each city grew rapidly between 1950 and 1970, as did the population of each city, as just noted. This is shown in table 28. The table indicates that in both regions the patterns of growth in agricultural production were similar to the patterns for urban population and employment growth, i.e., the rates of growth in both the 1950s and the 1960s were very high by ordinary standards, but the rates in the 1960s were considerably less than in the preceding decade. For both cities, therefore, the behavior of population and employment was generally consistent with the hypothesis that the growth of each city was a response to the growth of agricultural production in its hinterland.

Of course, general consistency between agricultural and urban growth rates does not of itself prove that urban growth was a response to the development of agriculture. A plausible case can be made for this hypothesis, however. The export theory of regional growth states that urban growth in the region is a response to expansion of regional exports. The first step, therefore, is to demonstrate that agricultural production in the Rio Yaqui and Costa de Hermosillo irrigation districts was primarily for export. In these districts in 1954, cotton production per head of population in Sonora was 9.9 times the per capita production in the rest of the country and wheat was 37.0 times as much. In 1960 these figures were 7.6 and 39.7,

Table 28. Agricultural Production and Population: Regions of Hermosillo and Ciudad Obregón

Hermosillo		Ciudad Obregón	
Agricultural production[a]		Agricultural production[b]	
Value	(millions of $)	Value	(millions of $)
1950-52	4.8	1952-54	23.1
1959-61	19.0	1959-61	42.7
1967-69	45.3	1967-69	74.5
Average annual growth (%)		Average annual growth (%)	
1950-52/1959-61	16.5	1952-54/1959-61	9.2
1959-61/1967-69	11.5	1959-61/1967-69	7.2
Population[c]		Population[c]	
Total	(000s)	Total	(000s)
1950	43.5	1950	31.0
1960	96.0	1960	68.0
1970	170.0	1970	114.0
Average annual growth (%)		Average annual growth (%)	
1950/1960	8.3	1950/1960	8.2
1960/1970	5.9	1960/1970	5.3
Economically active[d]	(000s)	Economically active[d]	(000s)
1950	13.0	1950	10.6
1960	25.9	1960	20.7
1970	44.5	1970	29.6
Average annual growth (%)		Average annual growth (%)	
1950/1960	7.1	1950/1960	6.9
1960/1970	5.6	1960/1970	3.6

[a] Production figures are for irrigation districts. Prepared under the supervision of Pierre Crosson by Manuel Garcia, student at the Center for Agricultural Economics, Postgraduate College of Agriculture, Chapingo, Mexico. Values expressed in average peso prices of 1964-66 converted to dollars at 12.5 pesos to the dollar. The basic data source is Estadistica Agricola, an annual publication of the Secretaria de Recursos Hidraulicos (SRH).

[b] Prepared under the supervision of Pierre Crosson by Gustavo Gonzales, student at the Center for Agricultural Economics, Postgraduate College of Agriculture, Chapingo, Mexico. Value expressed in average peso prices of 1964-66, converted to dollars at 12.5 pesos to the dollar. Source of data same as in footnote a.

[c] Figures are for cities of Hermosillo and C. Obregón 1950 and 1960 from Luis Unikel, "El proceso de urbanización en Mexico, Demografia y Economia 2(2) 1968; 1970 based on IX Censo General de Población, Resumen General, Dirección General de Estadistica (May 1970), Mexico, D.F.

[d] Data for 1950 and 1960 from Luis Unikel, "La población economicamente activa en Mexico y sus principles ciudades 1940-1960," Demografia y Economia (4) 1, 1970; 1970 estimated by Pierre Crosson on the assumption that the proportion of the economically active population in the population of the two cities was the same as in the state as a whole.

respectively. Cotton and wheat accounted for between 95 and 99 percent of total agricultural production in the two districts in 1954 and for 83 percent in 1960, as shown in table 29. If it is assumed that both in Sonora and the rest of the country all production of these commodities was either consumed locally or exported, and that per capita consumption of the commodities was the same in Sonora as in the rest of the country, then the per capita figures show that production of cotton and wheat in the regions of Ciudad Obregón and Hermosillo was much more export oriented than was production of these commodities in the rest of the country.

Table 29: Share of Cotton and Wheat in Total Value of Crop Production, Rio Yaqui and Costa de Hermosillo Irrigation Districts

	(percent)		
District	Cotton	Wheat	Both Crops
Rio Yaqui			
1954	60.2	34.7	94.9
1960	46.3	37.0	83.3
Costa de Hermosillo			
1954	65.8	34.1	99.9
1960	34.4	48.6	83.0

Sources: SRH, Estadistica Agrícola del Ciclo 1953-1954, Informe Estadistico no. 10, 1955, Mexico, D.F. and Estadistica Agrícola 1959-1960, Informe Estadistico no. 20, 1961, Mexico, D.F.

As a way of identifying export commodities, this is a conservative procedure for two reasons. It relates production in the Rio Yaqui and Hermosillo districts to population in the state of Sonora as a whole rather than to that of the cities of Hermosillo and Ciudad Obregón plus their hinterlands, which would be the appropriate measure. Hence the per capita production figures for the districts are biased downward for purposes of this test. Second, the procedure includes the possibility that the rest of the country was also exporting cotton and wheat (which it was). Hence ratios greater than one do not simply identify export crops in the two districts; rather they indicate that the region was even more specialized in production for export than the rest of the country. Indeed, per capita exports of cotton and wheat from the two districts (given the assumption about equal per capita consumption) were many multiples of per capita exports from the rest of the country.

Conclusive testing of the hypothesis that the growth of Hermosillo and Ciudad Obregon was a response to expansion of exports from their agricultural hinterlands would require detailed information about agricultural-urban relationships which, as noted above, does not exist. Somthing useful can be said nonetheless. The expansion of irrigation in the Rio Yaqui and Costa de Hermosillo districts after 1950 was accompanied by a rapid and large-scale incorporation of new inputs into the agricultural production function. Data on the growth in consumption of these inputs in the two irrigation districts are not available; however, since the districts accounted for about 70 percent of irrigated land and crop production in Sonora, statewide data give a good indication of changes in the two districts. The state data show that between 1950 and 1960, the area under irrigation increased by about 95 percent, the number of tractors increased by about 160 percent, fertilizer consumption grew by more than 500 percent, and the farm labor forces increased by about 50 percent, well in excess of the natural rate of growth.[69] While data on consumption of pesticides in

[69] Whitney Hicks, "Agricultural Exports and Economic Development: An Application of the Staple Theory to Sonora, Mexico," unpublished paper.

the state are not available, it is known that the quantity increased rapidly, particularly in cotton production, the dominant crop in the state. Wheat production was stimulated also by the rapid adoption of new seed varieties developed by the Center for Agricultural Research for the Northwest, an agency of the federal government located near Ciudad Obregón. As a result of these developments, agriculture in Sonora had achieved a high level of technological sophistication by the late 1960s. This is shown by the data in table 30 for the Hermosillo and Rio Yaqui districts.

Table 30. Fertilizer Use and Mechanization in Irrigation Districts of Mexico, by Region

Region or Nation	Percent of land using fertilizers[a]	Percent of land which was[b]		
		Totally mechanized	Partially mechanized	Not mechanized
Nation	68.3	47.2	36.5	16.2
Pacific North	82.3	77.6	20.5	1.9
Costa de Hermosillo	100.0	100.0	--	--
Rio Yaqui	67.8	98.6	1.4	--
North	52.0	34.1	50.1	15.9
Central	56.2	5.6	63.7	30.8
South	29.5	0.0	36.4	63.6

Source: SRH, La Mecanización Agrícola en los Distritos de Riego, Ciclo 1966-67, Informe Estadistico no. 41, Oct. 1968, Mexico, D.F. and El Uso de los Fertilizantes en los Distritos de Riego, Ciclo 1967-1968, Informe Estadistico no. 42, July 1969, Mexico, D.F.

[a]Crop year 1967-1968

[b]Crop year 1966-1967.

The growth of agricultural production thus was accompanied by a large increase in purchases of farm inputs. There also was significant growth in processing activities in and around Ciudad Obregón and Hermosillo. There is no information on the amounts of income and employment generated in the two cities by suppliers of farm inputs or by processors of farm commodities. However, even casual visual inspection indicates large numbers of vendors of farm machinery and parts, fertilizers, pesticides, seed, and feed, as well as cotton gins and flour mills in each city, suggesting that these activities must contribute significantly to income and employment. Silos' study of the Obregón area buttresses this conclusion. Referring to the early 1960s, Silos asserts that "the marketing structure of the Yaqui Valley is an integration of farm credit with marketing of agricultural products and the purchase of agricultural supplies."[70] The principal institutions engaged in marketing farm products were two public corporations; two federal agricultural banks; three credit unions; three private flour mills; two oilseed processors; ten cotton gins; and a large number of merchants, wholesalers, and middlemen. The range of services offered to farmers by these agencies included the extension of short- and medium-term credits and the purchase, storage, processing, and transport of goods. While some agricultural products were consumed in the region, most were shipped to other parts of the country and some were exported (out of the country). With respect to the supply of farm inputs, Silos asserts that

[70] Jose Silos, "The Yaqui Valley of Mexico: Its Agricultural Development, Resource Utilization and Economic Potential," unpublished Ph.D. thesis (Cornell University, Ithaca, N.Y., 1968) p. 74.

"most of the agricultural supplies and equipment are manufactured outside of the Yaqui Valley; however, there are a large number of commercial houses that distribute farm machinery and equipment, fertilizer, insecticides, and improved seeds."[71] While some of these marketing and supply activities are located outside Ciudad Obregón--a number of the agricultural processors for example--most of them are found in the city or its immediate environs and all of them contribute to the level of urban economic activity in the valley.

There can be little doubt that the production and processing effects of agricultural development in the two irrigation districts had significant impact on the growth of Ciudad Obregón and Hermosillo. In addition, the growth of net farm income must also have been important. This can be inferred from the extraordinarily large increase in agricultural production in the two irrigation districts. In the fifteen years from 1952-54 to 1967-69 the volume of production in the Rio Yaqui district grew by a factor of more than 3.2. In the Costa de Hermosillo the growth factor was 9.9 for the period 1950-52 to 1968-69. Net factor income--payments to land, labor, management, and capital--did not increase as rapidly as this because the share of purchased inputs in gross production rose. Nonetheless, the growth in gross production was so large that net factor income must have increased rapidly even if its share of gross output fell.

---

[71] Ibid.

Of course, not all of the increase in net agricultural income in the two irrigation districts would have been spent in the cities of Hermosillo and Ciudad Obregón. Some of it would have been used to directly purchase goods and services outside the region (i.e., bypassing local retailers and wholesalers), for "foreign" travel, and to accumulate savings in Mexico City and other financial centers. Nonetheless, a substantial proportion of the increase in agricultural income must have been spent in the two principal cities. Hence, while the income effect of increased agricultural production on local urban growth cannot be measured, there is every reason to believe it was significant.

We have no data showing quantitatively the contribution of the expansion and diversification of economic activity in Hermosillo and Ciudad Obregón to the growth of agricultural productivity in their respective hinterlands. There can be no question, however, that agriculture productivity increased substantially, nor can there be any doubt that this could not have occurred without the expansion of agricultural credit, input and marketing services provided by the two cities.

Conclusion

The material presented, interpreted in the light of economic base theory, suggests strongly that the expansion of irrigation in Sonora not only permitted rapid expansion of agricultural production but that it also indirectly induced fast economic growth in urban areas of the state. As part of the growth process, population in the state, and especially in the two principal urban areas, grew substantially in excess of natural rates of increase, indicating significant in-migration from other parts of

Mexico. The relationship between agricultural and urban development, however, was reciprocal: while the expansion of agriculture stimulated demand for urban goods and services, the resulting growth and increasing specialization of the urban economy facilitated technological advance and rising productivity in agriculture. The experience of Sonora, therefore, is generally consistent with the hypothesis established in chapter 5 of interdependence between agricultural and urban growth in a given state and with the implications of the hypothesis for state-to-city migration explored in chapter 6.

Chapter 8

CONCLUSION

At the beginning of this study we said that the study would be addressed to two questions: (1) Can public policy as applied to rural regions be expected to affect the rural-to-urban migration flow in Mexico and, if so, what are the more effective instruments for implementing such a policy? (2) Can such instruments be applied to alter the distribution of migrants as between Mexico City and other smaller cities?

We believe that the results of the study demonstrate that the answer to both questions is positive. Programs for the modernization of agriculture which increase the incomes of farmers by increasing their productivity will slow migration from rural to urban areas. And the stimulus given by agricultural modernization and higher farm income to growth of regional urban centers will divert some of the migratory flow from Mexico City to those centers.

We focused on the effects of large-scale investments in irrigation as the cutting edge of agricultural modernization, but comparable investments in drainage and flood control could have the same impact in increasing the productivity of labor and other resources employed in agriculture. There is considerable potential in Mexico for additional expansion of irrigation, drainage and flood control. Consequently, investments in these projects can be viewed as instruments of current urbanization policy in Mexico, although their main justification, of course, would be their impact in raising agricultural production and income.

While our results support this position, they also indicate that the effectiveness of these instruments in affecting urbanization is limited. The migrant stock variable was by far the most powerful factor determining the amount and direction of state-to-city migration between 1960 and 1970. The size of the migrant stock in each city cannot be much affected by policy except over the very long term--probably several decades. Thus for practical purposes the dominant element in state-to-city migration is beyond the reach of policy.

In addition the effect on migration of even massive investments in agricultural development is small in relation to total migration. Moreover, rural-urban income differentials are large and will remain so for a long time whatever measures are taken for agricultural development. Consequently, people will continue to have strong incentives to move from rural to urban areas. Finally, our analysis suggested that it is expectations of the *growth* of agricultural income which was most important in deterring migration. This indicates that a policy to slow migration by increasing agricultural income would have to be a sustained one to establish the expectation of further increase, and that the effect in slowing migration would be deferred, probably for some years.

The effectiveness of agricultural development as an instrument of urbanization policy is limited also because it conflicts in this respect with other policies also serving important development objectives. Our analysis showed, for example, that policies to increase the literacy of rural people would stimulate migration from rural to urban areas. Yet increased literacy is an imperative of development policy. Similarly, Mexico has a deep and long-standing commitment to the *ejido* program of land reform. Our analysis showed that states with a large percentage of land in *ejidos* on average had lower agricultural incomes per worker than states where *ejidos* were less important. Pursuit of the *ejido* program, therefore, might conflict with programs to raise agricultural income as a means of slowing migration. Our point, needless to say, is not that pursuit of the *ejido* program should slacken. Its long and still active life in Mexico indicates that the program serves deeply held social values in that country. The point is simply that as instruments of urbanization policy, programs designed to increase agricultural income may be constrained by the *ejido* program.

We do not argue that investment in modernization of agriculture is the only instrument of urbanization policy that Mexico ought to consider. We believe, although we do not claim to have demonstrated in this study, that such investments would be more effective than any others in slowing the rate of rural-to-urban migration. We do not believe, however, that they necessarily would be most effective in affecting the *direction* of migration as among alternative urban centers of destination. Our analysis shows that the effect of agricultural modernization on this aspect of migration is indirect. That is, the expansion of modern agriculture stimulates the demand for urban supplied goods and services, many of which can be most efficiently supplied in or near the area of agricultural production. Our analysis also shows, however, that alteration of relative wages among cities will affect migrant choices among cities. This suggests that policies which directly affect the demand for labor in cities can affect the direction of migration among those cities. Direct public investment in job creating activities in selected cities would serve this purpose, as would differential tax policies designed to stimulate private investment in those cities. Such policies in fact are employed in Mexico, at least in part because of their effect in diverting migration from Mexico City to secondary urban centers. The Border Industrialization Program, designed to stimulate manufacturing and other activity along the border with the United States, is an example.

In summary, programs to stimulate the modernization of agriculture can serve as instruments of urbanization policy in Mexico, but on any scale consistent with other development objectives their effect will be small in relation to total migration flows. Consequently, these migration effects are unlikely to weigh heavily in judging the desirability of programs for agricultural modernization. The primary criteria for such judgments likely will be the direct effects of the programs on such things as agricultural income and employment, food supply and prices, and balance of payments.

The programs will be most effective in slowing the overall rate of rural-to-urban migration, their effect on the <u>direction</u> of flow being indirect and smaller. The direction of flow probably can be influenced more strongly by programs which directly change the relative rates of increase of job opportunities among cities.

Appendix

Sources of Data Used in Chapters 3-5

Mexico: Inversión Publica Federal. Secretaria de la Presidencia, 1964. Mexico, D.F. Data for federal government expenditures in 1949-1953, by state, for health and education. Used to calculate GHEPP and GEDPP. (See chapter 4, p. 66).

Esso road map of Mexico. Data for highway distances between the principal city of each state and each of the 36 cities with 50,000 or more population in 1960. Used as SCDIS. (See chapter 4, p. 67).

Compendio Estatistico, 1951 and 1960, Dirección General de Estadistica, Mexico, D.F. Data for number of primary school personnel. Used to calculate EDTPP. (See chapter 3, p. 41).

Quinto Censo de Servicios (datos de 1965), Direccion General de Estadistica, Mexico, D.F. 1966. Data for wages and salaries and numbers of persons occupied in service industries, by state, in 1965. Used to calculate AUWAG. (See chapter 3, pp. 48-51).

Quinto Censo de Servicios (datos de 1965) por Zonas y Municipios de la Comisión General de Salarios Minimos (no publicada), Dirección General de Estadistica, Mexico, D.F. 1966. Data for wages and salaries and numbersof persons employed in each of the 36 cities with population of 50,000 or more in 1960. Used to calculate CWAGE60. (See chapter 4, p. 65).

Cuarto Censo Comercial y de Servicios (datos de 1955), Dirección General de estadistica, Mexico, D.F. Data for wages and salaries and numbers of persons occupied in 1955 in each of the 36 cities with population of 50,000 or more in 1960. Used to calculate CWAGE50. (See chapter 4, p. 65).

Censo Agricola, Gandero y Ejidal 1950 and 1960, Dirección General de Estadistica, Mexico, D.F. Data for value of agricultural production, number of tractors, total arable land, arable land in ejidos, and irrigated land in 1950 and 1960. Used in calculation of AGINC, TRCPH, IRRPL, and PABEJ. (See chapter 3, p. 41).

Censo General de Pobación 1950, 1960, and 1970, Dirección General de Estadistica, Mexico, D.F. Data for economically active population in agriculture, by state, 1950 and 1960; population between the ages of 10 and 29, by state, 1950 and 1960; percent of state population literate, age 6 or more, 1950 and 1960; total state economically active population in nonagricultural activities, 1960; migrant stock in each of the 36 cities, 1960; migration from states to each of the 36 cities, 1960 to 1970. Used in calculation of AGINC and PPLIT (chapter 3, p. 41); also SCMIG, MIGST, SLAFO, and PPYNG (chapter 4, pp. 65, 67-68); also PCSUR (chapter 5, p. 106).

Luis Unikel, "El proceso de urbanizacion en Mexico: Distribucion y crecimiento de la poblacion urbana," Demografia y Economia, vol. II, no. 2, 1968. Data on natural rate of population growth, 1940-1950 and 1950-1960, in 36 cities with population of 50,000 or more in 1960. Used in calculation of CLMPR. (Chapter 4, p. 67).

_____, "La poblacion economicamente activa en Mexico y sus principales ciudades, 1940-1960," Demografia y Economia, vol. IV, no. 1, 1970. Data used to represent CLAFO (chapter 4, p. 67); also in calculation of CLMPR (chapter 4, p. 67).

El Colegio de Mexico, 5 percent sample of residents in each of the 36 cities in 1960 who had been born in states other than the one in which the city is located, to determine date of arrival in the city. Data used by us to measure migration to the city from 1950 to 1960. (SCMIG in chapter 4, p. 65).

For Product Safety Concerns and Information please contact our EU
representative GPSR@taylorandfrancis.com
Taylor & Francis Verlag GmbH, Kaufingerstraße 24, 80331 München, Germany

www.ingramcontent.com/pod-product-compliance
Ingram Content Group UK Ltd.
Pitfield, Milton Keynes, MK11 3LW, UK
UKHW051833180425
457613UK00022B/1238

* 9 7 8 1 1 3 8 1 9 5 5 8 5 *